Engineering condition monitoring

Practice, methods and applications

Engineering condition monitoring
Practice, methods and applications

Editor: Dr Ron Barron

University of Strathclyde, Glasgow

© Addison Wesley Longman 1996

Addison Wesley Longman Limited
Edinburgh Gate
Harlow
Essex CM20 2JE
England

and Associated Companies throughout the World.

Published in the United States of America by Addison Wesley Longman Inc.,
New York.

Cover printed by The Riverside Printing Co. Ltd., Reading, UK
Text design by Sally Grover Castle, Reading
Illustrations by Chartwell Illustrators, Croydon, UK
Typeset by Techset Composition, Salisbury, UK
Printed and bound by T J Press, Padstow, UK

First printed 1996

ISBN 0–582-24656-3

British Library Cataloguing-in-Publication Data

A catalogue record for this book is available from the British Library

Library of Congress Cataloging-in-Publication Data is available

Contents

Contributors

All the contributors are or were employed in teaching and research at the University of Strathclyde, Glasgow.

Ronald Barron   ARCST, PhD, CEng, MIMechE. Senior lecturer in mechanical engineering; main interests are machinery dynamics, including condition monitoring studies, chaotic motion of engineering systems and active control of vehicle suspensions. Has edited all the contributions, written Chapters 1, 6, 7 and 12 completely, and supplied material for Chapters 4, 5, 8, 10 and 11.

James Brown   BSc, PhD, ARCST, RCST. Former reader and visiting professor in mechanical engineering; main areas of expertise are general dynamic analysis, design and instrumentation; still involved as a consultant in the field of refrigeration engineering. Has made a contribution to Chapter 5.

Ian A. Craighhead   BSc, PhD. Senior lecturer in mechanical engineering; main interests are the dynamics of suspensions and vehicles and machinery dynamics. Has contributed to Chapters 3, 4, 9, 10 and 11 and is the inventor of the Condition Monitoring Game.

John S. Fleming   BSc, PhD, CEng, FIMechE, MInstR. Senior lecturer in mechanical engineering; main interests are the dynamic and thermodynamic characteristics of helical screw compressors, refrigeration and the storage and transport of energy. Has made a major contribution to Chapter 8 and has contributed case studies for Chapter 11.

William Kennedy   BSc, PhD, CEng, MIMechE. Retired senior lecturer in mechanical engineering; main interest was machinery dynamics. Has written the majority of Chapter 4.

Rueben F. MacLean   MSc, PhD, ARCST, CEng, MIMechE. Retired senior lecturer in mechanical engineering; still actively involved as a consultant in condition monitoring. Has made contributions to Chapters 9 and 10.

Matthew R. Thomas   BEng, MSc, PhD. Former lecturer in mechanical engineering; main interests are machine condition monitoring and diagnostics, especially techniques, applications and management. Has written Chapter 2 completely, has written the majority of Chapter 10 and has contributed to Chapters 3 and 9.

Acknowledgements

All the contributors wish to accord their grateful thanks to the excellent work done by Caroline Houston, Mildred Troube and Marie Stewart in organising and typing the manuscript. The original diagrams were produced to a high standard by Sheena Nelson, Aileen Petrie, Patricia McKay and Isobel Mungall, and the contributors thank them for their work.

Introduction

1.1 Aims and objectives

It is now generally accepted that industrial machinery has a high capital cost and its efficient use depends on having low operating and maintenance costs. To comply with this requirement, condition monitoring of machinery and processes has become established throughout industry. Its methods have produced considerable savings by reducing unplanned outage of machinery, reducing downtimes for repair and improving reliability and safety. To capitalise on these advantages, the methods involved should be well documented so that up-to-date information is easily available to practitioners.

For a company to get full benefit from condition monitoring, its staff must be trained in a wide variety of engineering methods and procedures, so this book aims to provide essential information for training the following people:

- Technical directors and managers of plants who wish to introduce condition monitoring to help improve the efficiency and economic performance of a process.

- Practising engineers, at all levels of responsibility, who wish to become acquainted with the complete process of condition monitoring.

and for the use of:

- University and college lecturers who provide material for undergraduate and postgraduate courses.

The book gives up-to-date information for people who already use condition monitoring and for those considering its introduction. Besides this, it should help to coordinate the various subject areas. From the early techniques of some 30 years ago, using elementary methods and practical know-how, work progressed through sophisticated digital instrumentation for data collection and analysis on to the complex computer equipment and software of today. Practitioners need to have considerable knowledge to cope with the many strategies available.

This book aims to assemble that knowledge into a coherent framework. The task was narrowed slightly by considering the condition monitoring of rotating machinery, where most contributors have their expertise.

Furthermore, the contributors believe that vibration monitoring is the most extensively used in-house technique, so the main emphasis has been placed on this method in the text; other important techniques are summarised where appropriate.

1.2 Brief history

Condition monitoring has been developed extensively over a period of approximately 35 years. As a crude guide, condition monitoring has paralleled developments in electronic instruments, transducers, computers and software; it is now almost completely automated.

From 1960 to the mid 1970s simple practical methods were used, along with a careful watch on a machine's behaviour, often reinforced by frequent maintenance. Elementary instruments were sometimes used to measure and record the variables on which maintenance decisions were based. This required highly skilled and experienced maintenance staff to ensure efficient operation and to avoid catastrophic failures.

During the 1970s there were developments in analogue instrumentation and mainframe computers. Analogue instrumentation became popular in the form of portable vibration measuring and recording meters, FM tape recorders and frequency analysers. Many of these instruments were heavy and cumbersome but they did provide increased accuracy in recording and analysing variables. If a company had access to a mainframe computer, data would be kept in store so that maintenance strategies could be developed.

Although some digital instruments were available during the early 1970s, significant developments took place during the late 1970s and early 1980s due to the availability of the microprocessor. Circuits could be miniaturised, reducing the dimensions and weight of instruments and allowing data to be handled at high speeds. Onboard microprocessors gave instruments the ability to capture data, analyse it via a suitable algorithm, then store and display the information. A very significant feature of frequency analysis was efficient computation of the fast Fourier transform and the ability to store data for future decisions. This was often assisted by the range of minicomputers which emerged around that time. Long-term data storage became an accepted practice.

From the mid 1980s onwards the developments have been associated with the desktop computer, its interfaced equipment and software. Many manufacturers have produced hand-held instruments for the instant measurement, recording and analysis of variables; information is often available from the instrument on a component or machine condition. This makes it easier to decide upon maintenance strategies so the decision maker requires a lower level of skill. Indeed some companies have developed suites of software which allow the whole process of condition monitoring to be carried out automatically, giving a complete service for measurement, analysis and problem diagnosis followed by a maintenance strategy. The ultimate form of this approach is an expert system which eventually may lead to a fully automated condition monitoring system.

1.3   Structure of the book

Chapters 2, 3 and 4 contain introductory material. Chapters 5 to 10 contain the detailed information required for specific condition monitoring strategies. Chapter 11 presents nine case studies.

Chapter 2 surveys the background and gives an overview of the steps in condition monitoring. Advantages and alternatives are discussed.

Chapter 3 contains brief descriptions of vibration analysis, oil/debris analysis, manual inspections, current monitoring, conductivity testing, perfomance thermal and corrosion monitoring.

Because vibration monitoring is the most widely applied method, Chapter 4 deals with the essential theory of the vibration of machinery. Systems with single and multiple degrees of freedom are discussed under free and forced vibration conditions, and the effects of damping are described. Various sources of vibration are discussed, in particular rotational and reciprocating unbalance, seismic vibration, transverse vibration, whirling of beams and shafts and self-excited vibration. Special topics are also considered, such as vibration in the presence of dry friction, shaft hysteresis and vortex shedding in fluids.

Chapter 5, the first of the more detailed chapters, considers the main requirements for an efficient vibration measurement system. The next step is to gather the data, so Chapter 6 describes basic electrical and electronic principles and circuits, elementary analogue amplifiers and filters, vibration signal acquisition and conditioning.

Chapter 7, a natural extension to Chapter 6, surveys the well-established methods of vibration signal analysis, particularly discrete methods and digital instrumentation methods. Since there are a number of new and advanced techniques in this area, a fuller reference section has been provided.

If vibration is a main indicator of problems with a machine, the significant sources of vibration have to be catalogued. Chapter 8 provides information on two important sources of vibration: rotational and reciprocating unbalance. It describes the basic principles of both sources in mass/elastic systems with multiple degrees of freedom, along with methods of producing conditions which are acceptable in practice for efficient operation.

Most machines provide the power to drive other parts of a system, therefore, they use gears and shafts in bearings. Chapter 9 deals with vibration in bearings and gears. It summarises the sources and types of vibration in bearings and gears, derives an elementary theory for bearing frequency analysis and gives examples of this effect in ball/roller bearings, angular contact bearings and fluid-filled bearings; oil whirl is also considered. An elementary theory of gear vibration is used to highlight the importance of sideband frequencies in gearbox vibration. Trending and frequency analysis are briefly discussed.

The reader should now be ready to apply vibration condition monitoring to machinery, so Chapter 10 gives practical guidance on the setting up and running of a condition monitoring system. The topics include deciding what to monitor, selecting the appropriate technique, taking meaningful measurements and diagnosing faults. Practical guidance is also given on the use of condition monitoring procedures, machinery commissioning tests, periodic and malfunction tests and machine acceptance testing; Campbell and waterfall diagrams are discussed.

The nine case studies in Chapter 11 give a flavour of some practical applications. They range from elementary waveform analysis to the condition monitoring of a diesel engine/generator set. Other case studies look at coal-handling and ceiling tile plants, rotational balance studies, compressor vibration, and there is also a novel condition monitoring game.

The bibliography contains up-to-date reference material collected from about 1980 onwards. It deals with condition monitoring and related topics catalogued as follows:

- general machinery

- power plants and turbomachinery

- bearings and gears

- textbooks, standards and industry publications

There has been a great deal of literature on the subject of condition monitoring. The contributors would like to acknowledge their appreciation of other work in this field. Although the impetus to write this book came from a series of lectures presented by two of the contributors in 1988 and sponsored by Schlumberger Instruments, a great deal of inspiration and help was gleaned from the following major texts:

Angelo M. (1987). *Vibration monitoring of machines. Bruel & Kjaer Tech. Rev.* No. 1., 1–36.

Braun S. *et al.* (1986). *Mechanical Signature Analysis: Theory and Applications*. London: Academic Press.

Cempel C. (1991). *Vibroacoustic Condition Monitoring*. Chichester: Ellis Horwood.

Collacott R.A. (1979). *Vibration Monitoring and Diagnosis: Techniques for Cost-effective Plant Maintenance*. London: George Goodwin.

Wowk V. (1991). *Machinery Vibration: Measurement and Analysis*. New York: McGraw-Hill.

Inspiration also came from papers published in *Conference Proceedings of Condition Monitoring and Diagnostic Engineering Management* (COMADEM) between 1989 and 1994.

2

Condition monitoring: the basics

This chapter aims to explain how condition monitoring works and where it can be applied. The techniques used to acquire information about machine condition are dealt with in Chapter 3.

2.1   Historical background

The growth in condition monitoring systems can be considered from two angles: an industrial imperative towards effective maintenance and technological improvements to measuring equipment and accompanying software.

The demand for condition monitoring systems has increased as companies have tried to minimise the consequences of machine failures, and to utilise existing maintenance resources more effectively. Some of the factors which have increased this demand include

- increased quality expectations reflected in product liability legislation

- increased automation to improve profitability and maintain competitiveness

- increased safety and reliability expectations reinforced by legislation

- increased cost of maintenance due to labour and material costs

The supply of more cost-effective condition monitoring tools to satisfy industrial demand has been made possible through technological advances such as

- reduced costs of instrumentation

- increased capability of instrumentation, such as data presentation and storage

- improved data storage, using low-cost computer systems

- faster and more effective data analysis, using specialist software

Demand creation

Mechanisation and automation changed dramatically during the Industrial Revolution and has continued increasing to the present day. Before the Industrial Revolution the method of production mainly relied on manual labour, with few or no tools. Any tools that existed were extremely simple and of a sturdy design. The need for maintenance was minimal; failures were replaced, worn-out tools repaired. Obvious examples are horse-drawn carts, ploughs and even ships; they made use of simple materials such as wood, leather and small quantities of ironwork, which could easily be replaced or remanufactured.

From the Industrial Revolution until the more recent technological revolution, manufacturing moved away from manual labour for production and introduced machinery. A useful illustration of this process is the car industry. Originally, cars were hand-crafted, their body panels were formed and assembled manually. This gave way to a few simple body panels being formed using power-driven mechanical presses, with the bodies being manually assembled on continuous production lines.

The increased reliance on machines meant that any machinery failures had a direct effect on the profitability of companies, so maintenance was used to minimise the effects. The relative simplicity of machinery and the low price of labour meant that maintenance could be carried out by a combination of pre-planned tasks and repair when failures occurred.

The technological revolution has seen an acceleration in the use of expensive and complex machinery, allowing profitability to be realised through economies of scale. Today's car body plants use large numbers of body parts formed in automatic presses, and assembly takes place on a continuous production line worked by robots. Men and women are largely restricted to the supervision of machinery; only a few simple tasks involve direct labour.

The increased reliance on complex machinery has meant that any breakdowns have an even greater impact on profitability than before. This increased risk is due to factors such as loss of availability, cost of spares, cost of breakdown labour, cost of secondary damage and risk of injury to people and the environment. These pressures are reflected by recent legislation setting standards in areas such as product liability, health and safety, and pollution control.

The risk of machinery breakdowns has also been compounded because equipment and products are manufactured to tighter specifications, using minimal material to reduce manufacturing costs. This requires higher manufacturing tolerances to ensure reliable products of a high quality, as expected by equipment users. For example, car bodies are manufactured with higher tolerances today, using sheet steel of a thinner gauge but of higher quality, and people purchasing cars expect the paintwork to be totally free from blemishes.

The response of these pressures has been to demand maintenance systems able to minimise the risks of equipment failure. It is this feature which has been the driving force behind the development of maintenance systems based on the condition of equipment, rather than waiting until failures occur or replacing parts on a regular basis. As we shall see, the technological changes have also provided the means to produce condition monitoring systems in an effective, economic and efficient manner.

Systems development

The instruments and software required to carry out condition monitoring are defined by the need to take measurements, and the comparison of current and past readings to assess machine condition. A simple example of this process is the monitoring of oil consumption in a car engine. The oil level is measured on a regular basis and oil is added as required. Normally, a mental note is made of the oil consumption, and any increase in consumption over time warns of the possible development of a fault in the engine. In practice, instrumentation is used to collect the information, which is permanently stored to provide a history of readings.

Recent technological advances mean that the data collection instruments are now smaller and cheaper, they include more features and use less power. The shift from 1950s valve technology to present-day semiconductor design illustrates how the size of each component has decreased dramatically in size; and each smaller component is itself much more powerful. For example, a large number of bulky valves have been replaced by one small integrated chip.

These changes in size, cost and performance have increased the potential size of the market by providing cheaper and more effective tools for condition monitoring. For example, the price of low-cost portable frequency analysers fell by around 75% between 1980 and 1990. Equipment that once seemed marginal now looks far more attractive, particularly for smaller industrial companies.

In the past, the main methods of storing data used paper-based systems. They were labour intensive because records regularly had to be made out, then analysed. These shortcomings have been significantly reduced since the advent of low-cost desktop computers which can store large amounts of data and which have provided the platform for analytical software. Again, the development of low-cost data storage and analysis systems has increased the potential market for condition monitoring.

Furthermore, the development of software analysis tools has also improved the effectiveness of condition monitoring by providing greater access to expertise.

The instrumentation is now simpler to operate, and problems can be analysed with the aid of expert systems and knowledge-based systems. In contrast, early condition monitoring systems relied on a few experts having specialist knowledge of the instrumentation and analysis techniques.

2.2 Key features and fundamental steps

This section describes the basis for effective condition monitoring (CM) and the processes involved. The basis of condition monitoring is dealt with by addressing the key features of an effective system, including

- links between cause and effect
- systems with sufficient response
- mechanisms for objective data assessment
- benefits outweighing costs
- data storage and review facilities

This is supported by a brief review of component failure patterns, which explains why condition monitoring is effective and how the rate of data collection is important. Then comes a description of the following basic steps:

- identifying critical systems
- selecting CM techniques
- setting baselines/alerts
- data collection
- data assessment
- fault diagnosis and repair
- system review

2.2.1 When to use condition monitoring

Condition monitoring is based on the recording of measurements to predict the condition of equipment. A simple example of this process is the regular visual inspection of car tyres to assess their condition. The inspections are used to warn of increasing and uneven wear, external damage or insufficient tread. Although the majority of failures are covered, unexpected failures, such as running over a nail, are not.

The consequences of failure may be inconvenience, expense of repair, a possible accident due to total failure, or a fine for using a tyre with insufficient tread. Given the ease with which the checks can be made, the advantages far outweigh the costs of monitoring the tyres. Obviously, the readings would need to be taken often enough to identify possible faults, and to predict when a tyre would need replacing.

Although a simple example, it demonstrates that an effective condition monitoring system should possess the following features:

- A clear relationship must exist between the measurements being taken and the condition of the equipment.

- The monitoring system must respond quickly enough to provide warning of a deterioration in machine condition for appropriate action to be taken.

- The assessment of equipment condition must be made by comparing readings against existing measurements and/or against a predefined and absolute standard.

- The benefits of performing condition monitoring to predict equipment condition must outweigh the implementation and running costs.

- A system for measuring and recording data must exist to enable the condition of equipment to be predicted.

To understand how condition monitoring may be effective, it is worth examining the distribution of failures. This is illustrated by the 'bathtub' curve in Figure 2.1, which has three distinct stages: burn-in, random and wear-out. The exact shape of the figure will depend on the component; brake pads will have relatively few burn-in or random failures but a significant wear-out characteristic; bearings tend to have a high level of burn-in failures followed by randomly dominated failures.

Figure 2.1 illustrates that condition monitoring is applicable when failures are inherently random, and planned maintenance will only be of real benefit where there is a distinct wear-out characteristic. Using the curve, we can see that if planned maintenance was used, changing components at the end of the random

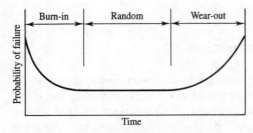

Figure 2.1   *The 'bathtub' curve.*

section of the graph, there would be a considerable number of failures between maintenance periods.

It is important that condition monitoring will be responsive enough to give sufficient warning of a degradation in performance. In terms of sampling, the rate of sampling should be at least 4 times greater than the time it takes for a fault to develop, with at least 4 or 5 readings being needed to establish a clear trend. For example, if a fault can develop over a minute, then the sampling rate should be every 15 seconds, which would only be achieved using a 'hard-wired' system. If, however, the time for a fault to develop is unknown, then the sampling rate can be approximated by ensuring at least 20 readings are taken between failures. So if a bearing fails every 2 years, then taking readings once per month should suffice.

2.2.2 How to implement condition monitoring

A condition monitoring system embodies several processes (Figure 2.2) grouped into two main areas: system set-up and review, and routine monitoring, assessment and diagnosis.

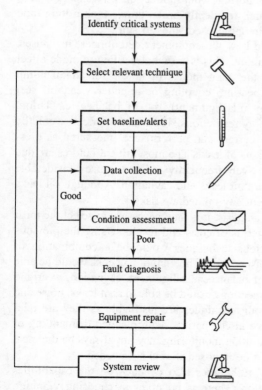

Figure 2.2   *The process of condition monitoring.*

System set-up and review

The setting up and review of a condition monitoring system can be summarised as follows:

- selecting the equipment to be considered for condition monitoring
- investigating how equipment fails
- finding the causes, effects and consequences of failure
- selecting a suitable condition monitoring technique
- deciding where and how often to take measurements
- taking baseline readings and setting alerts

The initial step is to identify equipment which would benefit from the application of condition monitoring. This may be achieved by examining all company equipment or by looking only at problem equipment. Eligible items will probably have a poor record of efficiency, availability, reliability and safety, or repair and maintainability; this is likely to show up as increased costs. An extreme example of a candidate for condition monitoring would be a diesel generator in a lighthouse which used excessive fuel, frequently broke down, took a long time to repair, was expensive to maintain and caused ships to sink!

The next step is to understand how the equipment deteriorates, its causes, warning effects and criticality of failure. This is essentially a failure mode effect analysis (FMECA). For example, the main modes of failure for an oil pump supplying a gas turbine may be bearing, coupling or impeller wear, oil seal failures (which may be caused by misalignment), out of balance, cavitation, overload, lack of lubrication or supply restrictions. The warning effects would be changes in vibration, temperature, current and performance (measured as pressures or flows), as well as visible signs of leaks. The most critical failures are due to bearing failure, which occur relatively frequently and may cause considerable damage; the least critical failures are oil leaks and cavitation. Although oil leaks occur frequently, they do not present any immediate risks.

Having completed the failure analysis, it is then necessary to select the most effective monitoring technique in terms of costs and results. Using the previous oil-pump example, the most obvious techniques would be a combination of vibration monitoring and general inspections. The vibration checks would be able to predict failures caused by out of balance, misalignment, cavitation, wear and lack of lubrication. The general inspections would be able to spot leaks, noise and changes in pump performance. Both techniques are justifiable as they are relatively cheap but extremely effective in comparison with planned dismantling of the equipment. Remember that condition monitoring may not always be the most appropriate method and should not be imposed if it is not needed.

Definition of the monitoring techniques also depends on the capabilities and commitment of the organisation. There is no point in choosing vibration measurement if the relevant equipment or experience is unavailable. But having selected an unfamiliar technique, it is possible to define the equipment, staffing, administration and training required.

The final set-up stage involves identifying where and how often to take the measurements, collecting the baseline readings and setting the alarm levels. Continuing with the oil-pump example, vibration readings would need to be taken montly on all the motor and pump bearings, and perhaps on the pump casing too. Before taking a full set of measurements, the measurement points would be clearly marked on the pump to act as the baseline for future readings. Alarm and warning levels would then be set; the alarm levels would be referred to a known standard and the warning levels set near to the system's normal reading.

The review process involves repeating these stages in light of operating experience, perhaps covering more machines or substituting one technique for another. Possibly the most important part of reviewing a condition monitoring system is the adjustment of the alert levels to increase or decrease sensitivity. If the alerts are not sensitive enough, failures may occur which have not been predicted by the monitoring system. If the alerts are too sensitive, there may be a large number of false alerts. In each case, the poor quality of the alerts could bring the monitoring system into disrepute and reduce its effectiveness.

Routine monitoring, assessment and diagnosis

At the heart of condition monitoring lies the collection, storage and interpretation of data. The sources of information will depend on the monitoring techniques:

- local inspections
- local instrumentation
- process computers
- portable monitoring equipment
- permanent monitoring equipment

and the methods of assessing the data:

- level checking
- trending against time
- comparison with other data

Careful processing of large amounts of data may be time-consuming. It is therefore important to make collection and assessment as efficient as possible, using the fewest staff and incurring the lowest running costs.

The method of data collection may be a completely hard-wired system, a manual system or a hybrid system (e.g. a portable data collector). The choice of method will depend on the level of protection that the equipment warrants, and the cost of the method itself. To avoid a catastrophic failure, a gas turbine may require a permanent system which is expensive and needs a quick response; an online hard-wired system would be justifiable because the benefits would outweigh the costs. In contrast, a non-critical pump would only require occasional checks, so a manual system would be more appropriate.

The system used to record the condition monitoring data will depend not only on the equipment, but also on the scale of the monitoring operation. A paper-based system is simple but recording and analysis are extremely labour intensive. A dedicated computer system may start out more expensive, but it stores the data effectively and can analyse the information properly when used with specialist software.

Data assessment aims to detect any deterioration in the equipment condition; it needs to be performed each time a new set of measurements are taken. The new readings may be checked against absolute alert levels, compared with past readings to detect any deterioration, or compared with the readings of other similar equipment being run under similar operating conditions. In addition, the readings should be checked for validity to ensure that a false alert has not been generated. For example, a set of vibration readings taken on a pump will first be checked against the warning and alarm levels; if an alert is generated, graphs of all the pump readings will then be checked to see if there is any deterioration that warrants further investigation.

Having identified a problem, it is necessary to find the cause and to ensure that correct maintenance action is taken. This may involve specialist knowledge to analyse the existing data or to carry out more detailed checks. A given problem may have several causes, so diagnosis will often involve a process of elimination to find those that are relevant.

2.3   Advantages and disadvantages

Condition monitoring can provide several advantages over other maintenance strategies:

- improved availability of equipment
- reduced breakdown costs
- improved reliability and safety
- improved planning

But it also has disadvantages:

- only marginal benefits
- sometimes difficult to organise

A government-sponsored study found the following gains from condition monitoring:

- On average, it achieved savings of 1.2% on value-added output (i.e. sales − material costs).

- Two-thirds of savings were due to an increase in availability gained from improved shutdown performance.

- There was a 75% decrease in unscheduled shutdown time.

- One-third of savings were due to a decrease in maintenance costs gained from improved breakdown performance.

- There was a 50% decrease in breakdown labour costs.

The costs associated with introducing a condition monitoring system were as follows:

- setting-up costs: 1% of capital equipment cost
- equipment purchase: 40% of setting-up costs
- training and gaining operating experience: 60% of setting-up costs
- running costs: one-third of total gross savings

Improved availability

Improvements in equipment availability can occur for several reasons. Firstly, condition monitoring can reduce the amount of planned maintenance, since maintenance is done only when needed. Secondly, if the total number of unscheduled failures is reduced, there will be a corresponding reduction in the overall time that a machine is unavailable while defects are being repaired. Finally, since condition monitoring can reduce the possibility of secondary damage, by predicting the onset of a failure, this will mean that the average repair time of an unexpected failure should also be less.

Reduced breakdown costs

Fewer spares should be required, since the extent of secondary damage is likely to be reduced, and fewer spares may be used in any planned maintenance. Benefits may also arise from the ability of condition monitoring to predict many failures. This means that spares need only be purchased when required, which in turn can reduce the numbers kept in stock. In addition, by reducing the number of breakdowns, the staffing levels needed to cope with unexpected breakdowns may also be lower, along with associated administration costs.

Improved reliability and safety

Improvements in reliability will be achieved for two main reasons. Firstly, by using the predictive nature of condition monitoring, it is possible to remove or replace a piece of equipment before any serious consequence arises. For example, if the beginning of a defect were identified in an aero-engine, it could be substituted, thereby improving aircraft reliability and passenger safety. Secondly, reliability can be improved by reducing the number of faults introduced during scheduled maintenance. This improvement is based on the premise that a significant number of faults develop during the early stages of a component's life, and that mistakes can be made during any rework, which can lead to failures. Improvements in reliability may diminish the insurance risk assessment and lower premiums.

Improved planning

One of the side-effects of introducing condition monitoring is that it can improve maintenance and production planning. This is for two reasons. Firstly, the ability to predict the onset of failures ensures that the organisation of materials and staffing can be carried out in advance, and fitted into any existing schedules. Secondly, the reduction in unexpected failures should reduce the need to reschedule or cancel existing work.

Disadvantages

Condition monitoring does have several disadvantages. It is unlikely to be 100% effective; a prediction rate of 80% is far more realistic. There may be a temptation to introduce condition monitoring in areas where the benefits are marginal, such as on non-critical equipment. Data collection, data assessment and diagnosis will incur costs in equipment and staffing.

The difficulties should not be underestimated. Effective condition monitoring requires appropriate equipment and properly trained staff. There will be an initial period where mistakes will occur as operaters gain in confidence and experience. If existing maintenance systems are in place, some consideration will be needed to integrate existing with new condition monitoring techniques.

2.4   Alternatives to condition monitoring

The aim of this section is to consider alternatives to condition monitoring and to compare their strengths and weaknesses:

- Condition monitoring
    - + improves availability
    - + reduces breakdown costs
    - + improves reliability and safety
    - − benefits may be marginal
    - − can be difficult to organise
- Failure-based or default maintenance
    - + effective for simple repairs
    - + requires no preplanning
    - − disrupts existing plans
    - − consequences may be unacceptable
- Design-out maintenance
    - + permanent solution
    - + can eliminate unacceptable failures
    - − can be lengthy and costly to implement
- Planned or preventive maintenance
    - + highly structured
    - + preplanned
    - − interferes with availability
    - − can introduce faults
    - − does not maximise life of parts

The type of management systems required are also mentioned, but this does not address the necessary support systems present in any organisation, such as purchasing or spares.

Failure-based maintenance

Failure-based maintenance allows equipment to run until a breakdown occurs; it is therefore reactive. This is the simplest approach to maintenance and is very effective where the consequence of failure and cost of repair are both low, as in the failure of an office light-bulb. Failure-based maintenance can also be considered as the default maintenance action, since the possibility of an unexpected breakdown will always exist.

The drawbacks depend on the consequence of failure. Unexpected failures are disruptive to existing activities; they require plans to be altered and may cause production losses. The failure of a conveyor on a car body production line can cause a considerable loss of profit. Consequential damage can arise out of a simple failure; the failure of an oil pump can cause an engine to seize, which is expensive to repair. This is true also in the case of gas turbines. The effect on safety may be unacceptable; the failure of a fire sprinkler system would expose people, equipment and buildings to the risk of an uncontainable fire.

The system required to implement such an approach can be extremely simple. A defect report is completed and passed to the maintenance department. The maintenance department raise a defect work order defining the repair, assign the job a priority according to its urgency, assemble the necessary resources, repair the defect and provide information for use in the future.

Design-out maintenance

Design-out maintenance is an approach which improves the performance of equipment by changing the design. Its main advantage is that a permanent solution is found, or normally at least! If the solution is correctly applied, it may improve reliability and reduce repair costs. The approach should therefore be used where improvements in reliability are needed or where maintenance costs are excessive for a given type of failure.

Redesign may be achieved either at the manufacturing stage, by selecting different materials and components, or after installation, by choosing an alternative equipment manufacturer. For example, if bearing failure is the most common defect of a central-heating water pump from a given manufacturer, there are two possible actions. The manufacturer can alter the design of the pump by selecting a different bearing which will not fail so readily, or the heating service engineer can fit a more reliable pump from a different manufacturer when repairing the heating system. Each action would produce a permanent solution to the problem, cutting down on warranty work for the manufacturer and increasing availability of the water pump.

One of the major drawbacks to this approach is that a design change may take a long time to implement. For example, an equipment manufacturer may not receive sufficient information for many months or even years about the performance of the equipment on site. Having identified any problems, it then takes time for the problem to be designed out, for the new design to be tested, and for production equipment to be altered before the revised equipment is ready for the market.

Consider another illustration. A pump in a petrochemical plant is identified as being a problem area and the decision is made to use an alternative manufacturer's pump. First of all, a viable alternative will have to be found, demonstrably better than the existing pump but still able to meet safety regulations. Changes to existing pipework and instrumentation may then be required, themselves time-consuming and also in need of testing and certification once completed. It is worth remembering that these barriers are not impossible to overcome, though the ease with which the changes can be effected will largely depend on the culture of the organisation involved.

The nature of design-out maintenance means that changes are normally a one-off event to eliminate a problem completely. This means there are usually two phases of implementation. Firstly, any problems justifying a change in design need to be identified by reviewing either maintenance history or customer feedback. Secondly, the problem requires designing out by a design team followed by testing. Finally, the changes need to be put into effect.

Planned maintenance

Planned maintenance is a system of carrying out maintenance at fixed periods, irrespective of the condition of the equipment. It is extremely effective where the performance and condition of equipment is related to the passage of time, and where the maintenance tasks can be carried out simply and quickly. Its main advantage lies in simple organisation; the resource requirements and timing of the tasks are known in advance. Some simple examples of planned maintenance are the regular lubrication of pump bearings and the changing of air filters which consistently become clogged over a period of time.

The disadvantages of planned maintenance are due to the random nature of failures; due to the possibility of introducing faults into healthy equipment (burn in); and because equipment must normally be taken out of use (off-line). Consider a bearing with the survival characteristics in Figure 2.3, having an L10 life of 500 000 revolutions. If a fixed-time maintenance policy were adopted, all the bearings would be replaced at L10, say every 2 years. This would mean that 90% of the bearings removed would be in a perfectly acceptable condition, and also that 10% of bearings would have failed in the 2-year period. Replacing the bearings every 2 years would be a costly exercise and would incur a risk of introducing more faults; it would also stop production for the maintenance to be carried out. In addition, 10% of bearings would fail again over the next 2 years. These unplanned failures would disrupt production and maintenance activities. If existing resources were not sufficient to deal with the failures, some planned maintenance activities would not be carried out, because staff would be diverted to unplanned failures. This in turn would lead to a greater volume of failures,

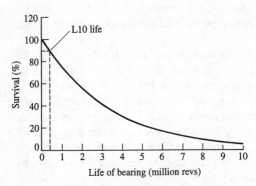

Figure 2.3   *Bearing survival characteristics.*

effectively producing a shift to failure-based maintenance, more commonly known as fire-fighting.

A number of side-issues exist as to what is planned maintenance and what isn't. Although the interpretation is fairly subjective, such questions at least deserve to be acknowledged. Firstly, inspections should really be considered as part of a condition monitoring approach. This is because no maintenance action is actually carried out unless an inspection, planned or unplanned, indicates it is necessary. Even though an inspection may be issued through a *planned* maintenance *system*, maintenance *action* is only carried out as a result of condition *monitoring*. Secondly, maintenance based on running time/number of cycles is a form of planned maintenance. This is because the maintenance action is triggered by machine running time/cycles irrespective of condition, and makes the assumption that running time/cycles are a good indicator of condition. In practice, it is customary to use a combination of elapsed time and run time, e.g. for car service intervals which are quoted as 'every 6 months/6000 miles, which ever is the sooner'.

Running a planned maintenance programme requires a well-organised and well-structured system, whether it uses paper or computers. Essentially, the system consists of a list of maintenance tasks which specify the job, the resources needed, the time to complete and the frequency. From this information, a plan is assembled which details the work to be done and its preferred timing. This plan is used to assess the resource requirements so they can be organised and ready when required, for the maintenance work. The final stage involves the issuing of jobs to maintenance personnel for completion. Although this is a simplistic model, it illustrates the need for defining maintenance tasks, for estimating and scheduling resources, and for controlling maintenance work.

3

Condition monitoring techniques

The aim of this chapter is to provide a brief description of the more common condition monitoring techniques, including

- vibration analysis
- oil/debris analysis
- manual inspections
- current monitoring
- conductivity testing
- performance monitoring
- thermal monitoring
- corrosion monitoring

In particular, the following areas are covered for each technique, then summarised in Table 3.1:

- A description of the more common analysis methods.
- An indication of the equipment to which the technique can be applied.
- An overview of typical failure modes that can be covered.

Table 3.1   *Summary of condition monitoring techniques.*

Technique	What to monitor	What to predict	Readings
Vibration	Gears	Imbalance	Acceleration
• Overall	Bearings (roller)	Looseness	Velocity
• Frequency	Couplings	Misalignment	Displacement
	Rotors	Wear	Spike energy
	Shafts	Poor lubrication	
		Cavitation	
		Clearances	
Oil/debris	Bearings (plain)	Wear	Composition
• Spectrography	Bearings (roller)	Fracture	Contaminants
• Ferrography	Gears	Contamination	
• Chip detection	Rotors	Degradation	
	Oil		
Inspections	Seals	Perishing	Temperature
• Internal	Cables	Wear	Leaks
• External	Switchgear	Overload	Noises
	Rotating equipment		Burning
Current	Motor windings	Poor insulator	Current
• Overall	Transformers	Worn brushes	
• Frequency		Failed rotors	
		Air gaps	
Conductivity	Motor windings	Overheating	Resistance
• Overall	Cables	Perishing	Capacitance
• Transient	Switchgear	Corrosion	
Performance	Filters	Blockages	Pressures
• Simple	Seals	Instrument drift	Flow rates
• Calculated	Motors	Perishing	Current
		Wear	Fuel usage
		Corrosion	
Thermal	Insulation	Perishing	Temperature
• Spot	Bearings	Overload	
• Thermal image	Coolant	Wear	
	Lubricant	Fracture	
	Switchgear	Chemical	
	Motors	reaction	

(continued)

Table 3.1 (*continued*)

Technique	What to monitor	What to predict	Readings
Corrosion	Structures	Chemical reaction	Dimensions
			Voltages
• Coupons	Pipelines		Resistance
• Electrical resistance	Vessels		
• Electrical potential			

3.1 Vibration monitoring

Vibration monitoring makes use of the vibrations generated by virtually all dynamic systems, e.g. rotating machinery. Individual components often generate distinctive vibration patterns which can indicate their condition or any change in their condition.

It is the ability to record and identify vibration 'signatures' which makes the technique so powerful for monitoring rotating machinery. And coupled with the relative ease and low cost of implementation, it explains why vibration monitoring is applied so extensively to rotating machinery, compared with other techniques. This section is merely an overview; more details are given in later chapters.

3.1.1 Techniques

Vibration analysis is normally applied by using transducers to measure acceleration, velocity or displacement. The choice largely depends on the frequencies being analysed:

- Acceleration covers frequencies from 0 up to and beyond 20 kHz.

- Velocity covers frequencies typically from 2 Hz to 2 kHz.

- Displacement, a measure of absolute position, covers frequencies from 0 up to 200 Hz.

The signals are normally processed and stored using a number of different methods. Some of the more common methods are described below.

Overall readings

Overall readings are obtained by taking the raw signal from the transducer circuitry and obtaining the peak, peak-to-peak or RMS values of the signal (like a
sound level meter on a hi-fi), they can then be recorded. This approach is simple
but tends to be insensitive to large changes in the amplitude of particular frequencies, although they make up only a small part of the overall signal. Overall
readings are often used in vibration standards, such as BS 4675, to indicate
machine condition.

Spike energy

This is a specialist technique developed by the manufacturer IRD; it is similar to
high-frequency detection (HfD from Endevco). The approach used is to pass the
incoming acceleration signal through an analogue filter, then to measure the
resulting amplitude within a frequency range, typically 5–60 kHz. It is effective
for problems such as poor/failed lubrication in bearings, problems which generate a broad range of frequencies.

Kurtosis and shock pulse methods

These methods attempt to indicate the spikiness of a signal by emphasising the
peaks in relation to the overall acceleration signal; they are commonly used in the
fault diagnosis of bearings. Kurtosis is based on a statistical approach and
measures $\text{amplitude}^4/\text{RMS}^4$. Shock pulse was developed by SPM; it records the
amplitude of the spikes along with the amplitude of the background 'carpet' level.
This information is then trended and compared in order to determine the type of
fault and the degree of severity.

Spectrum analysis

Spectrum analysis takes the incoming signal and breaks it into its individual
frequencies by using either analogue filters or a software process called Fourier
analysis (analogous to the human ear selecting the noise of individual instruments
within an orchestra). This process is extremely powerful and is used extensively
for trending and diagnosis; it relies on the ability to link particular frequencies to
particular components, such as bearings or gears. However, spectra generate large
volumes of information which require expert staff or software to interpret them.

Envelope detection

This specialised technique is applied in condition monitoring when background vibrations have to be suppressed, thus limiting the monitoring to the appropriate frequency ranges. In general, applying envelope detection to a signal ensures, to some degree, that the structural properties of a system, the system resonances, do not mask the vibration measurements under which monitor the condition of the machine. Enveloping achieves this by eliminating low-frequency vibration, then taking the frequency plot of the 'enveloped' signal (using a signal follower).

Cepstrum analysis

Cepstrum analysis is another specialised technique; it is used mainly in the analysis of signals that comprise a wavelet and one or more echoes, which may overlap. An advantage of measuring the cepstrum for machinery vibration is the ability to identify periodicity in a spectrum, such as harmonics or sidebands. Cepstrum analysis is useful for detecting and separating families of sideband frequencies in a gearbox spectrum, frequencies that indicate faults.

3.1.2 Applications

Many techniques exist which can be used either singly or in combination to identify vibration patterns. The components typically monitored include

- gears
- bearings
- couplings
- rotors/shafts
- fan/pump assemblies

And the techniques are able to distinguish several failure causes, such as

- imbalance
- looseness
- misalignment
- wear
- poor lubrication

- damage/perishing
- aero/hydrodynamic forces
- cavitation

Finally, the response of a vibration monitoring system to the onset of failures will depend greatly on the instrumentation. The faster the response required, the more likely a hard-wired system will be needed rather than portable vibration data collectors. Where the consequence of failure is higher or where access is a problem, the easier it is to justify the cost of a hard-wired system. For example, on a gas-turbine assembly, the vibration monitoring equipment is likely to be built in, whereas the ancillary equipment (e.g. oil lubrication pumps) may be monitored using portable equipment.

3.2   Oil/debris monitoring

Oil/debris monitoring is an extremely effective tool for assessing the condition of the oil itself and the components with which the oil comes into contact. It is particularly useful in equipment where vibration analysis is difficult to carry out, perhaps where many moving components are grouped closely together, or where components are remote from possible transducer mounting points. It is therefore typically used in compressors, gas turbines, reciprocating engines, gearboxes and other low/variable speed equipment.

Although oil/debris monitoring is used widely, it is often more cost-effective for oil samples to be analysed by outside companies; lubricant suppliers often provide this service free of charge. As the on-site input is likely to be less than for vibration analysis, this book only presents an overview.

3.2.1 Background

The oil in any system is often required to perform a number of functions, such as to reduce friction, to cool components and to clean load-bearing surfaces. Over time, the oil is likely to degrade, losing its lubrication properties due to chemical breakdown and becoming contaminated by the ingress of coolants, fuels and other lubricants. In addition, there may be a build-up of particles in the system, perhaps caused by component wear or the generation of soot.

The properties of the oil can be monitored in a number of ways; acidity for oxidation, viscosity for lubrication, flashpoint for contamination, and chemical composition for chemical degradation. In general, they indicate the quality of the oil and whether it needs to be replaced; they also indicate the failure of components such as seals and heat exchangers, as well as components overheating.

In addition to the oil properties, the presence of wear particles in the oil can also be used to predict a number of faults by observing their size, quantity, shape and material composition. These particles may be caused by wear (metal against metal), ingress (failure of filters) or corrosion of components; they may also be a by-product of reactions such as overheating (e.g. soot).

The chemical analysis of particles can often identify particular components which are failing (e.g. copper due to failure of bronze bearings/bushes), whereas shape often indicates the mechanism of the failure (e.g. production of loops of swarf due to abrasion). Combined with the quantity and size of the particles, therefore, this information not only identifies faults, but also the rate at which the fault is developing and how near to failure is the equipment.

3.2.2 Techniques

A number of techniques are available for analysing the condition of the oil and any wear particles that are present. The major techniques are summarised below.

Simple physical tests

A number of simple tests exist which can determine the lubrication properties and contamination of the oil. These tests typically consist of viscosity measurements, flashpoint and total acidity number (TAN). They are very general methods but viscosity and TAN together give an indication of oxidation, changes in flashpoint can indicate contamination, and changes in viscosity can indicate changes in the chemical structure of the oil.

Magnetic chip detectors

As the name implies, a magnetic plug is inserted into the lubrication system before the filter to pick up chips created due to components failing. Many of these plugs are self-sealing, which avoids loss of lubricant during inspection. The rate at which particles appear and the size of the particles give an indication of wear, but only where iron is present. Larger particles are often generated near to failure, so it is often more effective to use a continuous monitoring/protective system on equipment where the consequences of failure can be severe, such as bearings in gas turbines. However, cheaper systems simply rely on visual inspections and regular cleaning of the magnet by technicians.

Ferrography

Ferrography uses a strong magnetic field to separate particles according to their size. Two main methods are used. The first 'direct' method applies the magnetic field to a sample held in a glass tube. The larger particles move further than the smaller particles, and by using light emitters and sensors placed along the side (often just two), it is possible to estimate both the concentration and distribution of each particle size. The second and more accurate 'analytical' method applies the magnetic field to a slow flow of oil moving across a glass slide, the flow deposits the particles along the slide according to size. These particles can then be analysed, perhaps using high-power optics, to determine their size distribution and more important, their shape. Different failure mechanisms (e.g. abrasion, fatigue, corrosion) tend to generate distinct particle shapes. Normal wear generates small flat particles at low rates, with larger particles occurring more frequently at higher rates; 'loops' of metal are formed in cases of abrasion. It is therefore possible to assess whether normal wear patterns are occurring, or if a particular component is failing in a specific manner.

Spectrography

Spectrography is one of the main oil analysis techniques; it employs several different methods to identify concentrations of particular elements within the oil. These elements can be compared to an 'as new' sample for monitoring oil condition, monitoring the presence of contaminants and monitoring the presence and rate of generation of metal particles in the system. Typical wear indicators are copper, zinc, aluminium, iron, carbon and chromium; contamination indicators include silicon, sodium and boron. One of the limitations of spectrographic oil analysis (SOA) is that it is less effective when large particles are present in the oil. Some of the most common methods are atomic absorption spectroscopy (AA or AAS) and atomic emission spectroscopy (AES). In both cases, a sample of oil is taken and atomised, perhaps using a spark. The energies absorbed/emitted at particular wavelengths correspond to the presence of particular elements and the strength of each absorption/emission indicates the concentration. This information can then be trended and used to identify components undergoing excessive wear conditions. Refer to Section 11.4 for an example of oil analysis.

3.3 Manual inspections

In manual inspections the maintenance staff use their senses of sight, touch and hearing to make an assessment of the condition of equipment. The techniques used can vary extensively from the very simple to the involved, but all can be grouped into two main categories: internal and external.

Internal inspections can be carried out easily by dismantling the equipment, using inspection hatches or by inserting an optical probe (boroscope). Optical probes are preferable because they are non-intrusive and are less likely to introduce faults.

External inspections tend to be far more general; they are used to pick up obvious faults such as perishing of rubber components or loosening of bolts.

The types of components that can be monitored are limited only by access; any problems need to be deduced from the observations made. The following problems are typical of those that can be identified:

● Overheating: due to rubbing.

● Leaks: due to worn seals.

● Noise: due to vibration.

● Smells: due to overheating.

● Decay: due to perishing.

The strength of manual inspection lies in its low cost and simplicity; its frequency depends on the components being monitored. A gearbox may only need to be inspected for wear once a year using a boroscope, whereas a brake pad may require inspecting once a month.

Inspections often operate on a pass/fail basis instead of by trending. It is possible the equipment may fail at any time before the next inspection, and this could be crucial with safety equipment. An MOT one day does not guarantee roadworthiness on the next.

3.4   Current monitoring

Induction motors are used in industry for driving a vast range of machinery. The best maintenance strategy for motors depends largely on their size. It is feasible for many companies to carry spare motors and operate a breakdown or planned maintenance strategy, especially if the motors are small, inexpensive and drive non-critical machinery. This approach is very attractive if the motors on a site can be standardised so there are only a few varieties, reducing the range of spares.

For larger motors this approach becomes less attractive and the benefits of applying condition monitoring are hard to ignore. The motor is a fairly simple rotating machine and many of the vibration monitoring techniques described later can be applied to diagnose bearing faults, balancing problems and coupling problems. However, a far more comprehensive indication of the motor's electrical and mechanical condition can be obtained by monitoring the current passing through it. The signal can be picked up using a clamp-on ammeter or from the motor ammeter circuit if available. The signal can be analysed using Fourier techniques and it is possible to identify the following defects:

- broken rotor bars

- cracked rotor end-rings

- high-resistance joints in squirrel-cage windings

- casting porosities in aluminium die-cast rotors

- poor brazed joints in copper fabricated rotors

- rotor winding problems in slip-ring induction motors

- static and dynamic air-gap irregularities

- unbalanced magnetic forces

- mechanical unbalance

- bent shaft

- bearing irregularities

A typical commercial system is the Entek Motormonitor (Entek Scientific Corp., Cincinnati: On-Line, in Service, Diagnosis of Induction Motor Faults) which can be used on motors greater than 50 h.p. (37 kW). A PC and a spectrum analyser are needed to process the data and to provide diagnoses and recommendations.

3.5   Conductivity or insulation monitoring

Most electrical equipment relies on the effective insulation of electrical conductors. Deterioration and breakdown of insulation occurs naturally with time and can be accelerated if the equipment is operated outside it's design parameters or with faults present. Breakdown of the insulation can cause substantial consequential damage, usually produced by a fire or explosion. Monitoring is essential where breakdown is likely.

Large transformers are usually monitored regularly to detect deterioration in the oil/paper insulation which is widely used. Monitoring is usually based on sampling the oil and testing for dissolved gases produced by faults and insulation breakdown. Concentration measurements on hydrogen, hydrocarbons and carbon oxide gases enable faults to be detected. The type of fault can be defined using the following gas ratios:

- Acetylene/ethylene for discharge faults.

- Ethylene/ethane for high-temperature thermal faults.

- Methane/hydrogen for low-temperature thermal faults.

BS 5800 provides a guide for the interpretation of the analysis of gases in transformers and other oil-filled equipment in service.

Insulation deterioration in motor and alternator windings is easily detected by measuring coil resistance but this requires the machine to be taken out of service.

3.6   Performance monitoring

Performance monitoring uses process information to indicate the efficiency of equipment; any changes often indicate a deterioration in condition. A simple example is the monitoring of petrol consumption on a car; deterioration may indicate dirty inlet filters or poor ignition timing.

A large number of process readings can be monitored using this technique, but some of the more common parameters are

- differential pressure (filters)

- flow rates (pump delivery)

- absolute pressures (control valves)

- current (motors, generators)

- fuel consumption (diesel engines, gas turbines)

Again, the problems revealed by performance monitoring have many causes; some of the more obvious ones are

- blockages/clogging

- calibration

- perishing

- wear

- corrosion

The main advantages of performance monitoring are its cheapness, most of the instrumentation will often exist, and its simplicity. There are two main drawbacks. Many of the parameters depend on some other variable (such as load); this makes it more difficult to calculate a performance indicator. And the link between cause and effect is often less clear than in other techniques, which may then be required to pinpoint problems.

3.7   Thermal monitoring

Thermal monitoring is important for components which generate, transfer or store energy as heat. Typical items that are monitored are

- thermal insulation
- bearing housings
- coolant/lubricant
- heat exchangers
- electrical wiring
- fuse-boxes
- circuit boards
- transformers
- motors

And any temperature rises may be caused by

- damaged/perished thermal and electrical insulation
- incorrect electrical loading
- friction due to rubbing components
- exothermic chemical reactions

The methods used to monitor temperature fall into two categories: contact and non-contact. The main advantage of non-contact methods is that large areas can be surveyed quickly and at a distance; this may be beneficial where access is difficult (e.g. inside rotating machinery) or unsafe (e.g. high-voltage equipment).

3.7.1 Contact methods

Contact methods normally require the temperature measuring device to be placed on or within the surface of the component being assessed. This may be achieved using permanently installed devices or portable instruments.

Visual indicators

In addition to devices such as mercury thermometers, a number of simple paints, crayons and tapes are available which change colour as the temperature changes.

They are easy to install/apply and give a quick indication which can be observed easily by maintenance staff and operators. However, this method is better suited to 'OK/not OK' decisions than to trending.

Thermocouples/resistive devices

Thermocouples use wire of dissimilar metals (e.g. copper/constantan) which produce a voltage proportional to temperature in the linear range. Resistive devices measure resistance changes that occur with changes in temperature. These methods are often used in permanent temperature control and measurement systems. They are ideal for temperature trending. Portable instruments are also available in which the temperature sensor is fitted on the end of a probe.

3.7.2 Non-contact methods

The non-contact approach uses the principle that all bodies radiate energy in proportion to their temperature. It is also possible to relate the wavelength of the radiation to the temperature of the body. The main difficulty with this approach is that different materials emit different amounts of energy when at the same temperature (an effect known as emissivity). Fortunately, most opaque materials have an emissivity of between 0.8 and 0.95, which makes comparison straightforward. This means that when trending is used, the absolute temperature is not required. To obtain the temperature reading, most devices use a system to focus the energy from a source onto a sensor, whose reading can be processed and displayed as a temperature.

Pyrometer

A pyrometer is a device with sensors which accept a wide range of frequencies and have a simple focusing device (sometimes automatic). Pyrometers look quite similar to a single-lens reflex camera. They are often used to select point temperatures only, and are therefore of limited use compared with instruments that process and display the information as a two-dimensional picture.

Infrared

Infrared is often used in 'scanning' cameras which are able to produce a two-dimensional colour image. As the name suggests, the detectors use the radiation produced in the infrared part of the spectrum. Infrared cameras are extremely powerful diagnostic tools because they produce colour images which can pinpoint faults easily and quickly. But due to their cost, they are often hired directly or through a consultant when performing thermographic surveys.

3.8 Corrosion monitoring

The environment or process fluid used in plant operation can lead to corrosion of certain parts followed by gradual or catastrophic breakdown. The corrosion may lead to deterioration in performance of a machine, and this may be detected by performance or vibration monitoring methods, e.g. bearing failure or impeller unbalance.

Corrosion is usually a greater problem for structural components than for rotating machines, and a number of techniques are commonly used to determine its extent.

Electrical methods

Resistance A probe which consists of a thin wire is inserted into the process fluid. As the wire corrodes, it's cross-sectional area reduces and the electrical resistance increases. This can be measured with a suitable bridge circuit to give a continuous signal related to the rate of corrosion.

Polarisation Corrosion is an electrochemical reaction and it has been found that the slope of the potential/applied current curve is inversely proportional to the reaction rate on the electrode surface. This can be related to the rate of corrosion and probes are widely used to facilitate these measurements.

Potential The electrical potential between the plant being monitored and a reference electrode can be related to the rate of corrosion.

Coupons

Small pieces of the same material as the structure to be monitored are subjected to the same environment and are situated for convenient inspection and weighing at periodic intervals to trend the progress of the corrosion. Operating experience then enables the state of the monitored component to be estimated.

Ultrasonic testing

An ultrasonic pulse is fired from one side of the component and the time is measured for it to reflect from the other side and return. This indicates the thickness of the material and is a valuable method for monitoring pipe wall thickness and pressure vessel wall thickness.

Chemical analysis

Measurement of concentration of metal ions, pH, oxygen and hydrogen can give an indication of the corrosion process, depending on the materials involved.

Sentinel holes

Small holes drilled to specific depths in a pipe wall or similar component will leak when corrosion reaches the bottom of the hole. Single or multiple holes of varying depths can be used. The holes are often plugged once the leakage has been detected so that the plant may run until a suitable outage can be arranged.

Further information on these and other techniques for corrosion monitoring can be obtained from publications such as *Industrial Corrosion Monitoring* (HMSO 1978).

3.9   Other techniques

There are several other techniques besides those mentioned here, for example, ultrasonic testing and liquid dye penetrants. Although not dealt with in depth, they may be appropriate for special applications such as checking the main wing supports of an aircraft for fatigue and cracks.

4

Aspects of vibration

Vibration occurs in a wide range of engineering systems, particularly power generation and transmission, where the uneven transfer of energy (e.g. by an internal combustion engine) may be the prime cause, although both rotational and reciprocating unbalance may also play a large part. In general, vibration is seen as an undesirable by-product of the system, although there are certain instances where it is actively induced, e.g. vibratory feeder conveyors.

In the first instance, vibration may be considered as an interchange between two energy stores – potential energy (or strain energy) and kinetic energy – producing a cyclic or periodic motion of the system, i.e. motion which is repeated in equal intervals of time. For example, if a mechanical system possessing both mass and elasticity is displaced from its equilibrium position and then released, it will perform oscillatory motion.

The frequency of the resulting motion will correspond to one of the *natural frequencies* of the system – frequencies which the sysem will adopt when influenced only by the local parameters of mass and stiffness, in the absence of external effects. Associated with each natural frequency is a normal mode shape which shows the manner or movement of the system as it performs periodic motion.

Practical engineering systems tend to be very complex and extremely difficult to model analytically, but they can often be represented in a simpler manner by an ensemble of discrete elements consisting of masses, springs and dampers.

The vibratory motion of such a lumped-element system may then be described by the variables or coordinates associated with each element. The number of natural frequencies of a system, and consequently the number of modes in which it can vibrate, is equal to the number of degrees of freedom the system possesses. And the number of degrees of freedom within the system is defined by the number of independent coordinates required to describe the location of all mass elements comprising the system.

A system with a single degree of freedom will therefore require a single independent coordinate to fully specify the motion of the complete system; a system with three degrees of freedom will require three independent coordinates, and so on. Figure 4.1 shows some typical systems with a single degree of freedom.

If the rectilinear spring/mass system in Figure 4.1(a) is constrained to move in the vertical direction, it takes only the single translational coordinate y to completely describe the vibratory motion of the arrangement.

Similarly, if the pendulum shown in Figure 4.1(b) is constrained to oscillate in the vertical plane and the torsional pendulum shown in Figure 4.1(c) is constrained to oscillate only about the vertical axis, the motion of each may be defined by a single independent coordinate. Consequently, they may also be classified as systems with a single degree of freedom.

The number of masses may be increased to two, either as part of a rectilinear system, shown in Figure 4.2(a) and (b), or as part of an angular system, shown in Figure 4.2(c) and (d). A further coordinate is then required to fully specify the vibratory motion; each diagram in Figure 4.2 represents a system with two degrees of freedom having two normal modes of vibration, although in Figure 4.2(a) and (c) one of the modes is a 'rigid body' or zero-frequency mode, which is not usually of any practical interest.

The vibration of these systems can be visualised by plotting the displacement of each part of the system against its spatial position. For rotational or angular systems, more common in engineering, this plot of angular displacement versus position along the length of shaft will produce the broken line shown in Figure 4.2(c) and (d) – a normal elastic curve.

The assumption here is that the system is vibrating in a normal or principal mode associated with one or more of the natural frequencies of the system; if the shaft connecting the lumped masses or rotors is uniform, it will produce a straight-line variation of angular displacement. Then only the peak amplitudes of angular oscillation at the rotors need be known for the mode shape to be plotted. If the shaft or shafts between the rotors are not uniform (e.g. if they are stepped or if they are a combination of solid and hollow shafts), it is usually more convenient to resolve them into equivalent shafts of uniform diameter having the same angular stiffness as the actual combination, allowing the straight-line plot to be applied, as shown in Figure 4.3, and the location of the system's nodal points.

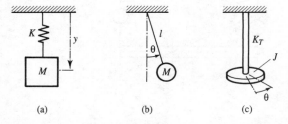

(a) (b) (c)

Figure 4.1 *Systems with a single degree of freedom: (a) rectilinear, (b) pendulum, (c) torsional pendulum.*

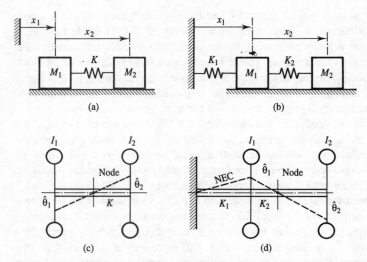

Figure 4.2 *Systems with two degrees of freedom: (a, b) rectilinear, (c, d) angular.*

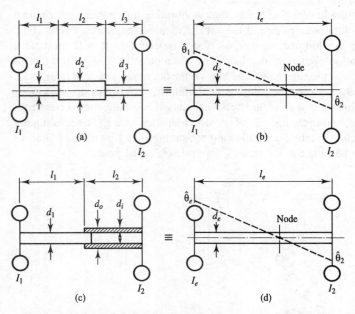

Figure 4.3 *Equivalent shafts: (a) stepped, (b) uniform diameter, (c) composite, (d) uniform diameter.*

Nodes or nodal points on an angular system represent plane sections of the connecting-shafts which are undisturbed by the vibration, i.e. places where the vibratory angular displacement is zero. Rotors that are immediately adjacent to nodes will vibrate so they are always in antiphase – at 180° to one another; strictly, this applies only to an undamped system.

The position of nodal points on systems that are subject to angular vibratory motion is important for both measurement and design, particularly if a geartrain forms part of the interconnecting link between the rotors.

Where a geartrain is present, it is often more convenient to resolve the geared system of multiple shafts into a single uniform shaft, as shown in Figure 4.4. The resulting modal plot will correspond to a straight-line relationship between angular displacement and longitudinal position along the shaft.

The relevant natural frequencies and associated modes of vibration grow more difficult to determine as a system becomes more complex (i.e. it has more degrees of freedom), so invariably they are analysed by computational techniques. Often, only the *lower* modes of vibration are of interest, particularly the fundamental mode. The fundamental will depend upon the number and distribution of the discrete masses comprising the systems as a whole. Figure 4.5 represents a four-cylinder diesel engine with equal rotors, angular inertias and shaft stiffnesses. It is a system with four degrees of freedom (including the zero-frequency or rigid-body mode). The three important normal modes of torsional oscillation associated with the respective natural frequencies are as shown.

If, however, this engine is coupled to a flywheel and is subsequently used to drive a boat's propeller, the modes of vibration will change considerably. It may be that only the fundamental mode of torsional oscillation will be of interest for this combination (Figure 4.6).

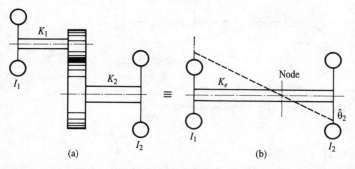

Figure 4.4 *Dynamically equivalent shafts: (a) geared shafts, (b) single uniform shaft.*

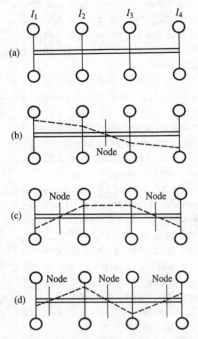

Figure 4.5   *Torsional modes of a four-cylinder diesel engine: (a) zeroth mode, (b) first mode, (c) second mode, (d) third mode.*

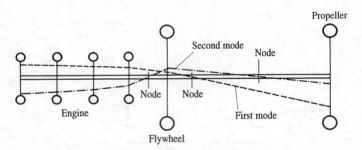

Figure 4.6   *Torsional modes of a combined system.*

4.1   Damped vibration

So far we have considered only free undamped vibration, where the principal frequencies and associated modes are free and undamped. The response at any point on a free undamped system (other than a node) will be as shown in Figure 4.7 – periodic motion with a constant maximum amplitude at the specific natural frequency and with no loss of energy. However, all practical systems are influ-

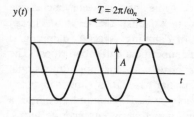

Figure 4.7 *Free harmonic motion.*

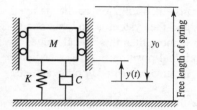

Figure 4.8 *Damped system with a single degree of freedom.*

enced by external forces, and the vibratory motion will diminish with time as the total energy of the system (kinetic and potential) is dissipated as heat or perhaps radiated as sound. Such systems are called *damped* systems; the mechanism of damping may be caused by friction in the air, in fluids or by friction between solids. Viscous damping can be modelled most simply when it is proportional to the vibratory velocity. Figure 4.8 shows a system with one degree of freedom where the presence of viscous damping is indicated by the oil-dashpot.

The damping action alters the natural frequency of the system and exponentially diminishes the amplitude of free vibration (Figure 4.9). The higher the level of damping, the more rapid the exponential decay of the waveform, i.e. the more quickly the vibration dies out.

Decay of transient vibrations can be used to establish the actual amount of viscous damping present within a system, even when that system is extremely complex. If the amplitude response of the system is recorded – on magnetic tape,

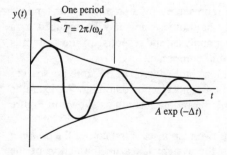

Figure 4.9 *Subcritical vibratory response of a damped system.*

as a chart or on a digital storage oscilloscope – the ratio between successive amplitudes will contain sufficient information to establish the level of energy dissipation within the system and the degree of damping. Viscous damping is usually described using three categories: critical, subcritical and supercritical. The mathematical solution of a second-order response with the non-dimensional viscous damping coefficient equal to 1 is a special case. It is shown in many standard texts (e.g. *Dynamics of Physical Systems* by R. H. Cannon) that, after a significant time, the system returns to a position which is, in engineering terms, the same as the static equilibrium position.

If the level of damping within the system is less than the critical value, on being disturbed the system will perform a transient oscillation at the damped natural frequency. This is subcritical damping and the system is underdamped.

If the level of damping within the system is more than the critical value, on being disturbed the system will return to its original position in a slow negative exponential manner, without vibrating. This is supercritical damping and the system is overdamped.

The actual amount of damping present within a system is expressed in the form of a non-dimensional ratio

$$\xi = \frac{\text{actual damping}}{\text{critical damping}}.$$

For critically damped systems $\xi = 1$; for underdamped systems $\xi < 1$ and for overdamped systems $\xi > 1$.

4.2 Forced vibration

Natural (unforced) vibration of a system is vibration in the absence of external influences besides the mass and the stiffness. Damped vibration is a special case of natural vibration under a local influence, here the influence is the damping or the energy-dissipating agent. However, in many practical systems, particularly rotordynamic systems, the local effects in addition have time-dependent forces applied directly or indirectly to the mass elements of the system. Time-dependent forces produce a marked change in the vibratory response of the system.

The time-dependent forces may be represented by a step, ramp or random input force. Periodic but not necessarily harmonic, they may be reduced to a series of harmonic functions using Fourier analysis; the response of the system is then obtained for each prominent harmonic component.

For simplicity, consider a system with a single degree of freedom possessing the physical characteristics of mass, elasticity and viscous damping. We examine its response to a force that varies in simple harmonic fashion as shown in Figure 4.10.

The complete response is the sum of two responses. First comes the *damped* or transient response; starting from rest, the system takes up the action of the applied force, causing it to vibrate at the damped natural frequency. Second

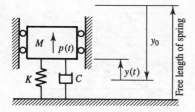

Figure 4.10   *Damped system with a single degree of freedom.*

comes the continuous or *steady-state* response; as time elapses, the transient is eliminated and the predominant feature is the steady-state response at a harmonic frequency equal to the frequency of the input force and with a maximum vibratory amplitude that depends on the system parameters (Figure 4.11).

The transient part of the response is of great importance where shock or impulsive forces act upon the system, but it may be neglected where long-term excitation of the system is likely to prevail. Long-term excitation of significant magnitude may induce fatigue failure of individual components or indeed of complete systems.

Of perhaps even greater importance in the steady-state response of the system is the frequency at which the forcing action takes place. Where this forcing frequency, ω, is much greater than the undamped natural frequency, ω_n, the vibratory amplitude of the system tends to zero, irrespective of the degree of damping present.

As ω approaches ω_n, the response of the system increases dramatically; and when there is zero damping ($\xi = 0$), the response is theoretically infinite. The condition where $\omega = \omega_n$ the undamped natural ω_n is known as *resonance*; the term is often misused to describe the condition when the vibratory amplitude reaches a maximum. This is not the case; Figure 4.12(a) shows that maximum amplitude occurs at a frequency lower than ω_n, i.e. at $f = (\omega/\omega_n) < 1$. But where the degree of damping is low, the error from this approximation is small.

Also important is the phase response – the relationship between the phase of the forcing action and the phase of the resulting displacement. Figure 4.12(b)

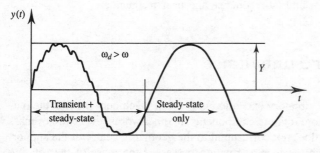

Figure 4.11   *Total response of the system in Figure 4.10.*

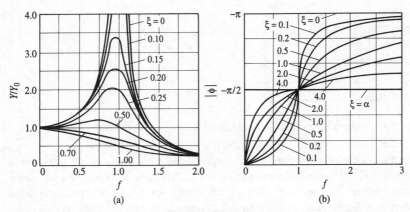

Figure 4.12 *(a) Amplitude and (b) phase response of the system in Figure 4.10.*

shows that, for any given frequency of excitation, the phase angle, ϕ, between the input force and the forced vibratory amplitude is constant, if the rest of the system parameters remain unchanged.

Resonance, at $f = (\omega/\omega_n) = 1$, always corresponds to a phase angle of $-\pi/2$ radians, irrespective of the level of damping. For the special case where no damping is present, the phase angle is zero for $f < 1$ and $\pm \pi$ for $f > 1$; in other words, the force and displacement are always in phase until resonance occurs; thereafter they are in antiphase. At resonance, the phase angle for zero damping is indeterminate.

All practical systems experience some degree of damping, and from Figure 4.12(b) we see that a more realistic value of ϕ ranges between 0 and $-\pi/2$ for $f < 1$ and between $-\pi/2$ and $-\pi$ for $f > 1$, depending on the level of damping. For any finite degree of damping, the condition of resonance always corresponds to $\phi = -\pi/2$ exactly.

At resonance, the reactive elastic and inertia forces of the system balance each other exactly, leaving the damping force as the only resistive force acting on the system. Thus the total resistance is a minimum at resonance, the impressed force and subsequent velocity of the system are in phase, producing large vibratory displacements both at resonance and near resonance.

4.3 Rotational unbalance

The primary source of vibratory excitation in rotating machinery is often the out-of-balance of one or more rotating components; in other words, the centre of mass of the rotating system does not correspond to the geometric centre of the support bearings through which the axis of rotation passes. If the mass of the rotating system is m and the distance between the axis of rotation and the centre of mass for the rotating system (i.e. the eccentricity) is e (Figure 4.13) then for a given rotational velocity, Ω rad s^{-1}, there will be a corresponding rotational force of magnitude $me\Omega^2$ applied to the shaft. The line of action of this force (the centrifugal force) will be from the axis of rotation radially outwards through the centre of mass of the rotating system.

Consequently, the cyclical force $\rho(t) = P \cos \omega t$, P constant – applied to the single mass/elastic system in Section 4.2 – will now be replaced by a corresponding cyclical force $\rho(t) = P \cos \omega t$ where $P = me\Omega^2$. If the total mass of the system (i.e. rotating and non-rotating) is M, then the lateral response of the system for varying values of frequency and levels of damping will be as shown in Figure 4.14.

Irrespective of the amount of damping present within the system, at rotational speeds much greater than the natural undamped frequency, i.e. $\Omega \gg \omega_n$ or $f = (\Omega/\omega_n) \gg 1$, the steady-state amplitude of its vibration tends to a constant, namely (me/M). And if reduction of the eccentricity is not possible, this amplitude can be kept very small using a large value of M, often achieved by attaching an auxiliary mass to the relevant part of a machine. Sometimes the auxiliary mass may be very similar to the mass of the machine itself, sometimes orders of

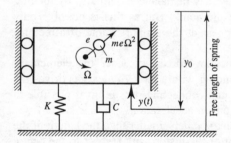

Figure 4.13   *Forcing by rotational unbalance.*

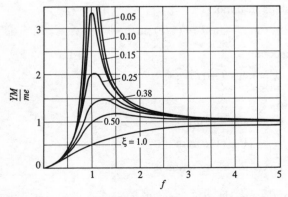

Figure 4.14   *Amplitude response of the system in Figure 4.13.*

magnitude greater. The combined mass may then be supported on relatively soft springs, which in turn will lower the natural frequency of the whole system.

For the special case where the impressed force is proportional to the square of the rotational speed, Figure 4.14 also shows that the frequency corresponding to the maximum amplitude condition is always greater than the resonant frequency. But as before, when the level of damping is low, there is hardly any difference between the resonant frequency and the frequency for maximum amplitude.

4.4 Basic principles of reciprocating unbalance

As well as studying rotary unbalance, it is important to consider the effects of inertia forces in reciprocators; they generally appear in automobile engines, railway locomotives, shipboard engines and reciprocating compressors. A crank and connecting-rod drive a piston in a fixed cylinder; this elementary planar mechanism produces complicated piston accelerations leading to inertia forces which are functions of the frequency components of the fundamental engine speed. When these forces are known, they may be balanced, partially, by rotating masses fixed in a position antiphase to the crank. But as this is possible only within practical limits, vibration may appear at the support frame of a reciprocating machine.

To assess the form of the reciprocating forces, consider the elementary mechanism shown in Figure 4.15. Expanding within the cylinder, gas drives the crank at a constant speed via the piston and connecting-rod, and it is possible to find an approximate expression for the piston acceleration.

Assuming that the length of the crank is much smaller than the length of the connecting-rod ($r \ll l$), the displacement of the piston with respect to the centre of rotation is

$$x = r \cos \theta + l \cos \phi \qquad (4.1)$$

With the following relationships in mind:

$$l/r = c \quad \text{and} \quad \sin \phi = (1/c) \sin \theta$$

the displacement equation becomes

$$x = r \cos \theta + rc \cos \phi \qquad (4.2)$$

Now

$$\cos \phi = (1 - \sin^2 \phi)^{1/2}$$
$$= [1 - (1/c^2) \sin^2 \theta]^{1/2}$$

The binomial expansion for $\cos \phi$ for a few terms is

$$1 - \frac{1}{2c^2} \sin^2 \theta - \frac{1}{8c^4} \sin^4 \theta - \frac{1}{16c^8} \sin^8 \theta - \ldots$$

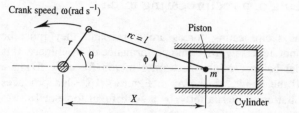

Figure 4.15 *Basic mechanism for a single reciprocator.*

Substituting this relationship into Equation 4.2 gives the displacement equation as

$$x = r\left\{\cos\theta + c\left[1 - \frac{1}{2c^2}\sin^2\theta - \frac{1}{8c^4}\sin^4\theta - \ldots\right]\right\} \qquad (4.3)$$

or

$$x = r\left\{\cos\theta + \left[c - \frac{1}{2c}\sin^2\theta - \frac{1}{8c^3}\sin^4\theta - \ldots\right]\right\} \qquad (4.4)$$

Differentiating Equation 4.4 with respect to time and setting $\dot{\theta} = \omega$, the collected terms produce the following equation for the velocity of the piston:

$$\dot{x} = -\omega r\left\{\sin\theta + \frac{1}{2c}\sin 2\theta + \frac{1}{8c^3}\left(\sin 2\theta - \frac{1}{2}\sin 4\theta\right) + \ldots\right\} \qquad (4.5)$$

To find the approximate acceleration of the piston, Equation 4.5 is differentiated with respect to time to give

$$\ddot{x} = -\omega r\left\{\cos\theta + \frac{1}{c}\cos 2\theta + \frac{1}{4c^3}(\cos 2\theta - \cos 4\theta) + \ldots\right\} \qquad (4.6)$$

Morrison and Crossland's *An Introduction to the Mechanics of Machines* may be consulted for a fuller derivation of Equations 4.5 and 4.6.

With a piston mass m kg, we get the following form for the inertial force transmitted to the frame via the crank bearing:

$$IF = -m\ddot{x} = -m\omega^2 r\left\{\cos\theta + \frac{1}{c}\cos 2\theta + \ldots\right\} \qquad (4.7)$$

The terms above the second-order component of the fundamental frequency quickly converge to zero, so Equation 4.7 adequately represents the form of the inertia force. But if the l/r is significant, several of the higher-order terms may prove to be important.

From a practical viewpoint, it will be prudent to have an added mass rotating at the crank speed, so only the first-order reciprocating force will be partially balanced. This feature is explored in the next section.

4.4.1 Partial balancing of a reciprocating force

If the magnitudes of the reciprocating forces are significant, they must be balanced so that they cannot adversely affect the performance of machinery. It is possible to provide partial balance of the first-order inertia force by fixing a rotating mass at 180° to the crank, as shown in Figure 4.16. This produces complete balance of the first-order inertia force by a centrifugal force at 180° to the crank.

Therefore

$$m\omega^2 r \cos \omega t = m_0 \omega^2 r_0 \, e^{j(\omega t + \pi)} \tag{4.8}$$

Equation 4.8 shows that the peak inertia force occurs at a crank angle of zero and the peak centrifugal force occurs at a crank angle of 180°. Although the magnitudes of the forces can be equated when the speed is constant, the phase difference has still to be considered.

Adding the forces in Equation 4.8 vectorially using complex number notation, gives

$$\frac{m\omega^2 r}{2}\left\{e^{j\theta} + e^{-j\theta}\right\} + \frac{m_0 \omega^2 r_0}{2} e^{j(\theta + \pi)} \tag{4.9}$$

Now since $mr = m_0 r_0$ and $e^{j\pi} = -1$, the sum (4.9) may be reduced to

$$mr\omega^2 \left\{\frac{e^{j\theta}}{2} + \frac{e^{-j\theta}}{2} - e^{j\theta}\right\} \tag{4.10}$$

In this equation the total force has a modulus of $|m\omega^2 r|$ and a phase variation of

$$\frac{1}{2}\left\{e^{-j\theta} - e^{j\theta}\right\} \quad \text{or} \quad -j\sin\theta$$

Coupled with (4.10), this expression suggests that the peak value of the force occurs at a crank angle of either $\pi/2$ or $3\pi/2$; in fact, at right angles to the direction of the inertia force.

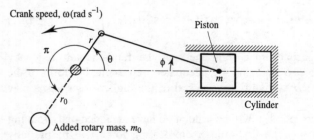

Figure 4.16 *Single reciprocator with rotary balancing.*

The effects of a reciprocating mass cannot therefore be completely balanced by a single rotating mass attached at 180° to the crank. To produce partial balance and to minimise the first-order inertia force, it is normal to assume that

$$m_0 r_0 = kmr$$

where the constant k is chosen to produce the minimum inertia force. Substituting this relationship in (4.10), the vector addition of the reciprocating force and the rotating force becomes

$$\frac{m\omega^2 r}{2}\{e^{j\theta} + e^{-j\theta}\} + km\omega^2 r e^{j(\theta+\pi)}$$

or

$$m\omega^2 r \left\{ \left(\frac{1}{2} - k\right)e^{j\theta} + \frac{1}{2}e^{-j\theta} \right\} \qquad (4.11)$$

When the results of this equation are plotted in the complex plane for suitable values of θ, the force values form a reverse-rotating ellipse (Figure 4.17). The significant values of force occur in the horizontal direction and at $\pm 90°$ to this axis – in the vertical direction. Therefore our attempt at partial balancing has only adjusted the inertia force; it has not been removed completely. The magnitude of the constant k critically dictates the value of the largest force and the direction in which it occurs.

To summarise, the first-order inertia force may be partially removed only by fitting a rotating balance-weight and is replaced by a reduced force in line with the piston or an increasing force with maximum values at angles $\pm \pi/2$.

Internal combustion engines normally contain more than one assembly of cranks, connecting-rods, pistons and cylinders. Indeed most commercial engines

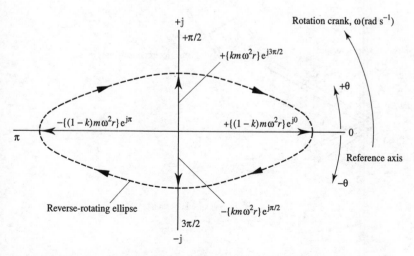

Figure 4.17 *Partial balancing of first-order inertial force.*

have multiple cylinders arranged in-line or in a V-bank. To find the inertia effects in commercial engines, it is helpful to derive a general expression for the reciprocating force in the basic mechanism (analysed in Section 4.4) then to adapt the analysis for multicylinder engines, both in-line and V-bank. The balancing of reciprocating effects in multicylinder engines is covered in Section 8.5.

4.4.2 A general expression for a reciprocating force

We have now developed an equation for the inertia force of a single-cylinder arrangement. Equation 4.7 can be used in the analysis of single-cylinder engines with horizontal or vertical motion of the piston, it can therefore be adapted to deal with in-line arrangements. We can also use the general analysis to find the inertial forces in a V-bank engine.

Consider the single-reciprocator arrangement in Figure 4.18, where the line of motion of the piston is at an angle of $-\alpha$ measured with respect to the vertical reference axis. From Equation 4.7, the general expression for the inertia force in a single reciprocator is

$$\text{Inertia Force } (IF) = C_n \cos n\theta \qquad (4.12)$$

where n is the harmonic order, here $n = 1, 2, 4, 6$, etc., C_n is the magnitude of the force and $\cos n\theta$ is the phase angle. Using this notation, the general expression for

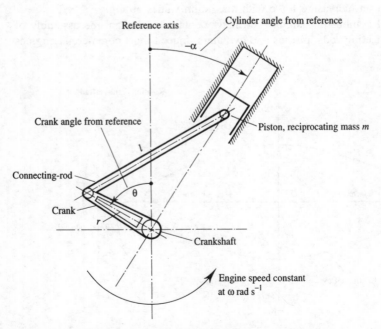

Figure 4.18 *Single reciprocator with inclined cylinder.*

the inertia force in line with the piston, up to the nth order component is

$$IF = C_n \cos n(\theta + \alpha) \tag{4.13}$$

or in line with the vertical axis:

$$IF = C_n \cos n(\theta + \alpha)e^{-j\alpha} \tag{4.14}$$

where again C_n is the magnitude, $\cos n(\theta + \alpha)$ is the phase angle and $e^{-j\alpha}$ is the spatial position of the piston.

Equation 4.13 can be modified to deal with in-line or V-bank arrangements before attempting to partially balance the inertia force/piston using an approach similar to Section 4.4.1.

4.5   Foundation force and transmissibility

The damping of a vibratory system transfers to the supporting structure some of the original force which set the system in motion. It does not matter how the force initiates the motion, some or maybe all of it will be transferred to the supporting structure. It is called the foundation or frame force and it should be kept to a minimum.

For simplicity, return to the system of Figure 4.10. It has a single degree of freedom and is acted upon by a harmonic force $\rho(t) = P \cos \omega t$. The transfer of this force to the support structure is described in Figure 4.19 using the *transmissibility ratio* (TR) where

$$TR = \frac{\text{Foundation force transmitted}}{\text{Applied force}}$$

As in Figure 4.12, the curves are plotted for various values of ξ. Notice that TR is less than unity only when $f > \sqrt{2}$, i.e. when the forcing frequency $\omega > \sqrt{2}\omega_n$. The damping reduces the maximum value of the force transmitted to the

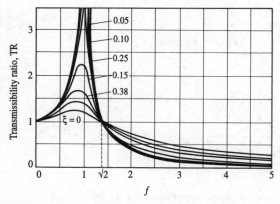

Figure 4.19   *Force transmitted to foundation.*

foundation only when $f > \sqrt{2}$; although above $f = \sqrt{2}$ the presence of damping actually increases the force transmitted. Beyond the point $f = \sqrt{2}$ there occurs what is called *vibration isolation*; this is somewhat misleading because the vibratory force is still present in the structure, albeit at a level lower than the excitation force.

In order to achieve maximum vibration isolation, i.e. minimum transmissibility, the undamped natural frequency of the system must be made as low as possible, using soft coil springs in the supporting structure, using additional mass (see Section 4.3) or by a combination of both.

4.6   Transverse vibration

Earlier sections have modelled the vibratory machines as discrete systems or lumped systems. Their constituent parts have been treated as idealised point masses or rigid bodies connected by massless elastic units. Such systems have finite degrees of freedom, a finite set of natural frequencies and a finite number of vibratory modes. Many practical problems can be analysed in this way, but in some real systems it may not be possible to separate mass and elasticity so easily. Shafts, beams, cables, plates and shells, all of them create difficulties for rigorous solution as discrete elements. They are better dealt with using distributed parameters, an alternative form of modelling where mass and elasticity are considered to be continuously distributed throughout the system.

The uniform shaft shown in Figure 4.20 is a system with distributed mass. It is capable of being excited in an infinite number of modes of lateral or transverse vibration, modes whose magnitude and form will be considerably influenced by the support bearings. In the simplest of cases, where the bearings are self-aligning, the normal modes associated with the natural undamped frequency of lateral shaft vibration will consist of a series of half-sine waves as shown in Figure 4.21.

Figure 4.20   *Uniform shaft in self-aligning bearings.*

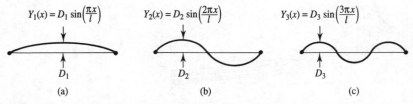

Figure 4.21   *Normal modes for transverse vibration of a uniform shaft: (a) n = 1, (b) n = 2, (c) n = 3.*

If, in addition to its uniform mass distribution, concentrated loads are attached to the shaft, the change in the total energy of the system (i.e. kinetic and strain energy) will alter the value of the natural frequencies. And depending upon their number and distribution, the loads may also affect the shape of each normal mode.

Instead of it being simply supported, suppose the shaft is a cantilever (i.e. overhanging the bearing) rigidly fixed at the bearing end, to prevent lateral movement, and free to move at the other end. This will alter the mode shapes considerably. The fixed/free boundary conditions also apply to individual blades in compressor/turbine arrangements, where the added influence of centrifugal loading can greatly influence the value of the natural frequencies. Figure 4.22 shows how the overall value of the natural frequency is influenced by two frequency components: ω_n^f due to normal transverse vibration, and ω_n^c due to the centrifugal effects.

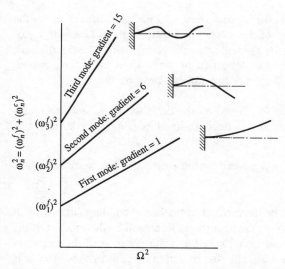

Figure 4.22 *Normal modes of a rotating cantilever beam.*

4.7 Whirling of shafts

So far we have defined the free and forced vibration of a system in terms of its natural frequencies and their corresponding normal modes. But an equally important phenomenon is *shaft whirling*. Although not strictly a vibration, it produces similar problems in rotating shafts, especially if they are thin and flexible. Shaft whirling occurs when the centre of mass of the shaft section does not coincide exactly with the geometric axis of the shaft; this may be caused by inhomogeneities in the shaft material or small deviations in geometry that lie within manufacturing tolerances. As the rotational speed of a defective shaft is gradually increased, a violent instability will eventually be encountered; apparently hinged at the bearings, the shaft will be deflected into a single bow as it whirls around like a skipping-rope.

By holding the rotational speed at this instability setting, the deflection is likely to become so large that the shaft eventually fails. But if, when increasing the rotational speed, the instability setting is rapidly passed through, the shaft will be deflected only briefly and will soon return to stable rotation, until the next instability is encountered. This instability will occur at a higher speed and will deflect the shaft into a double bow. Still higher speeds will produce triple bows, quadruple bows, and so on. Each speed at which an instability occurs is called a *critical speed* of whirling. Not only do instabilities occur on unloaded shafts, but also on shafts carrying concentrated loads such as rotors.

Let us consider the simplest case where a heavy rotor of mass M is attached to a thin shaft of negligible mass, as shown in Figure 4.23; this system is often called a *Jefficott rotor*.

Let BB be the line between the bearing centres, O the geometric centre of the shaft and G the centre of mass of the rotor. The centre of mass G and the geometric centre O will not normally coincide. Let the eccentricity e be the distance between O and G, and let r be the deflection of the shaft at the rotor position. Ignoring gravitational effects, there are two forces acting on the rotor:

- The elastic restoring force F_e offered by the deflected shaft tends to return the shaft to the equilibrium or unconstrained position, i.e. it tends to straighten the shaft.

- The centrifugal force F_c is produced by the rotation of the shaft and acts at G.

The elastic restoring force will be the product of the lateral bending stiffness K of the shaft and its radial deflection r; the centrifugal force will be the product of the rotor mass, the square of the rotational speed and the distance of the centre of mass from the line of the bearings. The line of action of each of these forces is shown in Figure 4.24. The system response for various rotational speeds, Ω, is shown in Figure 4.25.

In a similar fashion to the forced vibration of a single mass/elastic system, the deflection of the rotor/shaft arrangement builds up to extremely large levels

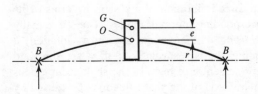

Figure 4.23   *Rotor spinning below critical speed.*

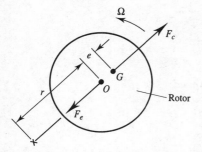

Figure 4.24   *Centrifugal and elastic forces acting on rotor.*

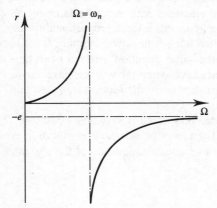

Figure 4.25   *Amplitude response of the rotor in Figure 4.24.*

around a critical speed Ω_c; this corresponds to the natural undamped frequency of lateral vibration of rotor and shaft, treating the shaft as a lateral spring. Indeed, the deflection would become infinite if it were not for the finite degree of damping that is present in all practical systems (hysteretic, viscous, aerodynamic, etc.).

The response is such that when the shaft is running below the critical speed Ω_c, both r and e have the same sign; the centre of mass G therefore lies further from the line BB than the geometric centre O (Figure 4.23). However, when Ω is greater than Ω_c, r and e are of opposite sign; G then lies between BB and O (Figure 4.26). As Ω is further increased beyond the critical condition, r approaches the value $-e$; G tends to a position coinciding with BB and the rotor rotates about the centre of mass with a high degree of stability.

In this analysis it was assumed that the bearings at BB were completely rigid, but many practical bearings, especially pedestal bearings, have a flexible supporting structure that is likely to influence the critical speed and the form shaft deflection (see Section 8.1).

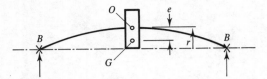

Figure 4.26   *Rotor spinning above critical speed.*

4.8   Self-excited vibration

In a self-excited vibration (SEV), the alternating force that sustains the motion is created or controlled by the motion itself; when the motion stops, the alternating force disappears. Another way of looking at the problem is by defining a common example of self-excited vibration as a free vibration with *negative* damping. A conventional system with positive damping generates a force proportional to the velocity of vibration and in the opposite direction. A negative damping force is also proportional to the velocity but has the same direction. Instead of diminishing the amplitudes of free vibration, negative damping will increase them. Positive or negative, the damping force vanishes when the motion stops, so this definition agrees with our first suggestion.

In most practical cases, the negative damping force is very small compared with the elastic and acceleration forces. Therefore the self-excited vibration is usually at a frequency which is very close to a natural frequency of the system. Consider some examples.

4.8.1 Dry friction

The characteristic dry friction curve can be the cause of self-excited vibration (SEV). The curve shown in Figure 4.27 is typical of the frictional force between two bodies as a function of velocity. The idealised system to the right of the curve consists of a mass restrained elastically sitting on a belt travelling at constant velocity. When an oscillation commences, the forces on the mass depend on the direction of the velocity. More work is put into the system during the second part of the motion than is lost during the first part; the motion builds up until non-linearities restrict its amplitude.

An example of this effect is the violin. A steady speed of the bow causes self-excitation of the strings and perhaps a pleasant sound! More often than not, the sound from this stick/slip phenomenon is not so pleasant. A torsional vibration of this type has been observed in ships' propeller shafts when rotating at very low speeds. The shafts are usually supported by one or two outboard bearings made of lignum vitae (guaiacum) or hard rubber, the bearings are normally water lubricated. At low speeds, no water film can form and the bearings

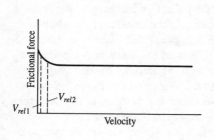

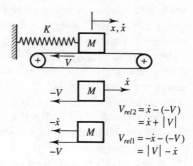

Figure 4.27 *SEV caused by dry friction.*

are 'dry', this causes a torsional vibration of the shaft at one of its natural frequencies, usually in the audible range.

4.8.2 Shaft hysteresis

Hysteresis is caused by a deviation from Hooke's stress–strain law and appears in most materials when they are subjected to alternating stresses. Instead of the familiar straight line, the material's behaviour is represented by an ellipse (Figure 4.28). For metals, the ellipse is hardly distinguishable from the straight line but its effects can be pronounced. The ellipse is more distinct for rubbers and other polymers.

For a component under alternating tension and compression, the stress–strain relationship is governed by a point moving around the ellipse in a clock-

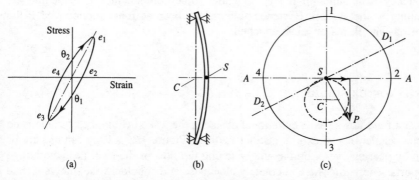

Figure 4.28 *Shaft hysteresis: (a) a straight line becomes an ellipse; (b) a vertical shaft in two bearings; (c) the centre of the shaft S describes a circle about the point C on the bearing centreline.*

wise direction, i.e. the strain lags behind the actual stress. Consider a vertical shaft in two bearings as shown in Figure 4.28(b). During the whirling motion, the centre of the shaft S describes a circle about the point C, which lies on the bearing centreline. Figure 4.28(c) shows a cross-section of the middle of the shaft with points 1, D_1, 2, 3, D_2, 4 lying on the shaft surface; the dotted circle is the path of S during the whirl. It is assumed that the rotation of the shaft and the whirling action are both clockwise. The shaft is bent and the line AA divides it into two parts so that the fibres of the shaft 'above' AA are elongated and those 'below' AA are compressed. The line AA is a line of zero strain, but because of the deviation from Hooke's law it does not coincide with the neutral line of stress.

Consider point 1 in Figure 4.28(c); this is a point of maximum elongation of the fibres, which will correspond with point e_1 on the stress–strain curve. In the course of shaft rotation, the fibres at point 1 go through the sequence e_1, e_2, S_1, e_3, e_4, S_2. And because the shaft is whirling, S and the line AA move around the dotted circle. If the speed of rotation is faster than the whirl, point 1 will gain on the line AA and move to 2, 3, etc.

If the speed of rotation is slower than the whirl, point 1 will move in the opposite sense, i.e. 4, 3, 2. For a rotation which is faster than the whirl, the point of zero stress e_2 lies between points 1 and 2, and e_4 lies between 3 and 4. The line D_1D_2 represents the line of zero stress; fibres 'above' this line are in tension; fibres 'below' this line are in compression. This stress system sets up an elastic force P with a component towards C and a component to the right, tending to drive the shaft in its path of whirl. There is therefore a self-excited whirl due to the hysteresis in the shaft. For the case when the rotation is slower than the whirl, point 1 moves anticlockwise and meets the point of zero stress e_2 before reaching the point of zero strain 4, therefore D_1 is in the top left quadrant of Figure 4.28(c) and D_2 in the bottom right. The component of P which is tangential to the whirl path is acting to oppose the motion and provides additional damping, therefore no whirl should occur.

Internal hysteresis of the shaft acts as damping on the whirl below the first natural frequency, and above this critical speed a self-excited whirl at the critical frequency may build up if the tangential force component can overcome the damping in the system. Hysteresis effects can become more pronounced if the shaft has shrink-fits or joints attached.

4.8.3 Vortex shedding

When a fluid flows past a cylindrical obstacle, the wake behind the obstacle is no longer regular, it will exhibit distinct vortices (Figure 4.29). They can usually be seen by peering over a bridge and watching the flow of the river near the bridge supports. The vortices are alternately clockwise and anticlockwise; they are shed from the cylinder in a regular manner and generate a sideways force on the cylinder. Experimental studies have suggested the following relationship between

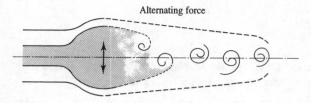

Figure 4.29   *Development of a vortex street.*

the frequency of vortex shedding F_s, the cylinder diameter D and the flow velocity U:

$$\frac{F_s D}{U} = 0.22$$

The quotient on the left-hand side is a dimensionless parameter called the Strouhal number; the relationship applies to a large range of flow conditions. The eddies are shed on alternate sides of the cylinder and they generate a harmonically varying force approximately equivalent to the force from the stagnation pressure of the flow. If the natural shedding frequency is near to a natural frequency of the cylinder, resonance occurs and may produce vibrations with a large amplitude. This type of motion has been observed on electrical transmission lines, submarine periscopes, industrial chimneys, bridges, pitot tubes, etc.

Many remedies have been developed; the Stockbridge damper, seen on power cables about 2 m from the insulators, is a short length of cable with a mass at each end and its centre clamped to the main cable. Vortex-induced vibration is absorbed by the masses and by twisting of the damper cable. Steel chimneys often have helical strakes fitted to break up the regular formation of vortices or damped guy-ropes to absorb chimney vibration. One of the most spectacular engineering failures ever to be filmed was the collapse of the Tacoma Narrows Bridge in 1940. Relatively low winds formed vortices at a frequency which coincided with a torsional mode of the structure, a more complicated effect than negative damping. The principal cause was aerodynamic flutter, an effect first observed on aircraft wings.

The motion was quite a tourist attraction and local drivers would often dare each other to cross the bridge as the oscillations grew more violent. Just a few months after its opening, the bridge finally collapsed. But redesigned to increase its torsional rigidity, the bridge has operated safely ever since. The modifications remove the troublesome structural mode from the vortex shedding and reduce the forces generated. Besides bridges, rotating machinery can also be prone to self-excited or unstable vibration; several mechanisms will be considered in Chapter 9.

References

Bishop R. E. D. and Johnson D. C. (1960). *Mechanics of Vibration*. Cambridge: Cambridge University Press. Reissued with Minor revisions 1979.

Cannon R. H. (1967). *Dynamics of Physical Systems*. New York: McGraw-Hill.

Den Hartog J. P. (1956). *Mechanical Vibrations* 4th edn. New York: McGraw-Hill. Reprinted by Dover, New York, 1985.

Gorman D. G. and Kennedy W. (1988). *Applied Solid Dynamics*. London: Butterworth.

Morrison J. L. M. and Crossland B. (1970). *An Introduction to the Mechanics of Machines*. London: Longman.

Newland D. E. (1989). *Mechanical Vibration Analysis and Computation*. London: Longman.

Rayleigh J. W. S. (1896). *The Theory of Sound* Vols 1 and 2, 2nd edn. London: Macmillan.

Thomson W. T. (1988). *Vibration Theory and Applications* 3rd edn. Englewood Cliffs NJ: Prentice Hall.

Timoshenko S. P., Young D. H. and Weaver W. J. (1974). *Vibration Problems in Engineering* 4th edn. New York: Wiley.

5

Vibration measurements

Vibration measurement, recording and analysis are now well-established procedures in machinery health monitoring. This chapter is the first of three chapters which summarise the main requirements for understanding

- a vibration measurement system
- acquisition of vibration signals with electronic equipment
- subsequent analysis and instrumentation options

5.1 Vibration measurement systems

A convenient and well-established approach to engineering system representation and analysis is the use of block diagrams (Cochin and Plass 1990). Block diagrams allow complicated engineering systems to be replaced by a block notation and specified input and output conditions, conditions which comply with the mathematical equation describing the dynamic behaviour of an engineering system.

The general form and stages of a measurement system are represented by the connected blocks to form the block diagram of Figure 5.1. It is assumed that no energy is lost in the whole process of measurement and result production. This feature cannot be achieved in practice so a 'good measuring system' will be one in which energy loss is a minimum. According to Figure 5.1, the main stages are

- primary sensing
- variable conversion

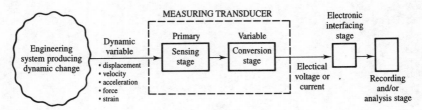

Figure 5.1   *Important stages in vibration measurements.*

- electrical interfacing
- recording and analysis

Primary sensing and variable conversion are normally carried out within the transducer, electrical interfacing involves separate electronic circuitry and recording/analysis requires specialist instruments.

To allow an adequate understanding of variable conversion and electrical interfacing, Chapter 6 covers certain elementary principles; it also gives some information on recording and analysis.

Primary sensing senses the parameter to be measured in a variety of ways (Doeblin 1990). Modern transducers come in two types: compression and planar shear; a good example is the piezoelectric accelerometer (see the Bruel & Kjaer Master Catalogue of Electronic Instruments, 1989, Naerum, Denmark). Schematic diagrams of both types are shown in Figure 5.2. The transducer operates as a seismic device, where a wafer of piezoelectric material is mounted as a mass/spring and damper system within a steel casing. When mounted on a vibrating surface, the piezoelectric material changes shape, producing a change of strain in the material and a change in electrical charge (or perhaps a change in voltage). The charge (or voltage) is proportional to the acceleration of the surface; it represents the variable being measured, and the conversion of the variable into

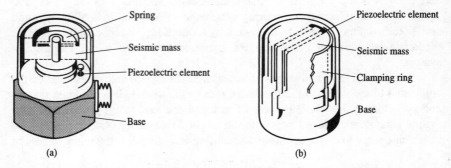

Figure 5.2   *Piezoelectric accelerometer: (a) centre-mounted compression, (b) planar shear. (Courtesy B&K Ltd)*

charge/voltage completes the second stage of the process. The representation of the variable is then electrically interfaced to recording/analysis equipment.

All the stages in vibration measurement are very important; they are summarised throughout the next few sections and accompanied by appropriate references collected at the end of the chapter. But first an introduction to the dynamic characteristics of some transducers.

5.1.1 Dynamic behaviour of the transducer system

The dynamic (or static) behaviour of the measurement system must be considered as a whole. This means that the influences on the dynamic behaviour of all the stages in vibration measurements have to be known to a suitable accuracy. A 'good' measuring system not only produces low energy loss but must also contain elements which give an acceptable dynamic behaviour.

The individual item/stage or the complete system is generally calibrated by the manufacturer and compared with standards (American National Standards Institute (1959); International Organisation for Standardisation (1990); British Standards Institution (1988). The dynamic behaviour of an individual element or complete measurement system may be described by a first- or second-order linear ordinary differential equation with suitable damping over an adequate frequency range; this is illustrated later in Figures 5.4 to 5.7.

5.1.2 Why measure vibration?

Modern plants containing rotating machinery have a high capital cost. As a general rule, the machines must be available for use as specified by the particular industry. This highlights the need for some method of monitoring a machine's 'health' while it is running. In this way the behaviour of the machine can be monitored throughout its economic life and maintenance procedures planned on long, short or emergency time periods.

If the condition of a machine deteriorates, the vibration associated with it will generally alter in a predictable way. By measuring and analysing the vibration of a machine, it may be possible to determine the nature and extent of the deterioration and thus predict how the machine will behave in the future.

5.2   Vibration monitoring and vibration analysis

Vibration monitoring and vibration analysis encompass a wide range of important mathematical topics. The mathematics required may seem rather off-putting, so

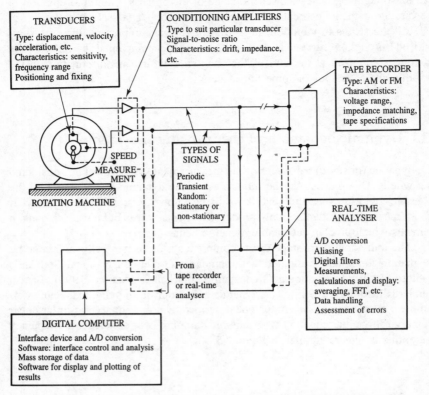

Figure 5.3   *Typical instrumentation for monitoring and vibration analysis.*

we have tried to interpret the results in a physical way to highlight the engineering applications and assist with their understanding.

By way of introduction, Figure 5.3 sets out the typical instruments and the principal topics of study. An engineer needs to be familiar with these topics before he or she can obtain reliable information on the health of a machine. Let's look at some of the options:

● Recording of vibration on a tape recorder for later analysis, or even just viewing on an oscilloscope.

● Recording and analysis of vibration by a real-time analyser.

● Recording and analysis of vibration on a dedicated digital computer.

To understand all of the options, the instrumentation system can be subdivided into the following six categories:

● vibration transducers

- signal-conditioning amplifiers for vibration transducers
- signal types
- instrumentation tape recorders
- real-time analysers
- dedicated digital computers for monitoring and analysis

Two themes run through them: signal acquisition and signal analysis. And the next section looks at transducers and instrumentation. The material may be widely applicable and is probably relevant to your transducers, but it remains important to consult the manufacturer's literature whenever you are making measurements.

5.3 Transducers and instrumentation

The acquisition and analysis of vibration signals depend on the quality of the measurement. It is of little value to perform highly sophisticated analytical techniques on suspect data. The choice and location of measuring transducers are of paramount importance in vibration investigations, especially for condition monitoring of installations.

Vibration transducers may be active or passive. Typical active transducers are quartz crystals and electrodynamic transducers; typical passive transducers are strain gauges and capacitance transducers. Active transducers could be described as self-powered, and passive transducers as powered from an external source. It follows that active transducers absorb all their energy from the vibrating system, whereas passive transducers can make more modest demands on the system.

Even non-contacting active transducers can absorb significant energy, and variable reluctance 'telephone earpiece' transducers can significantly affect the natural frequency of low-mass systems. For example, flapper-valve reeds used in refrigeration compressors experience frequency changes due to the variable stiffness imposed by alteration in the air-gap as the flapper vibrates.

Now that we have modern instruments, we no longer depend on high-energy active transducers, but they still have useful applications, particularly in hazardous environments. The quartz-crystal accelerometer is an active transducer.

5.4 Seismic transducers

Vibration measuring transducers are usually based on the seismic principle and may be used to monitor displacement, velocity or acceleration. The seismic principle is applied because most measurements are relative, e.g. movement of one body relative to another. In the absence of a fixed measuring 'table', an artificial or seismic table must be generated, usually by coupling the measuring stylus or location to a comparatively large mass using a light spring. For certain applications, e.g. gyrocompasses, a rapidly rotating angular inertia forms an extremely rigid seismic table.

Beginning with linear vibrations, three basic parameters are measured by transducers – displacement, velocity and acceleration. Perhaps the most common vibration measurement transducer is the quartz or barium titanate accelerometer, noteworthy for its high operating frequency range and its suitability for shock and vibration studies.

For lower frequency ranges, especially where low- or zero-frequency accelerations are to be monitored, the strain-gauge double-cantilever instrument or null balance (servo) type accelerometer can be employed. Displacement and velocity sensors are very similar to accelerometers; a displacement sensor measures relative displacement and a velocity sensor measures relative velocity.

All these instruments require some form of amplification or signal conditioning to accomplish signal acquisition then analysis. We shall divide the discussion into two parts:

- Characteristics and advantages/disadvantages of transducers (Section 5.4).

- Application of amplifiers and requirements for feeding recorders, etc. (Chapter 6).

Seismic transducers are most commonly adopted for vibration measurement, usually in the form of an accelerometer. The basic seismic transducer is shown in Figure 5.4; it consists of a mass supported by a spring in a casing which is fixed to the vibrating machinery; some damping is usually present. Certain transducers, particularly double-cantilever strain-gauged accelerometers, have a large amount of damping to extend their frequency response at high frequencies. It also helps to lower the transducer's tendency to 'ring' or vibrate under an impact or shock load. The amount of damping is carefully designed to preserve the proper operation of the instrument.

In simple terms, the mass tends to remain stationary and the relative motion between mass and vibrating body can be regarded as the absolute motion (relative to the earth). The term *seismic* is derived from the study of earthquakes, which first employed these devices.

At low frequencies, the mass tends to move with the vibrating body and the relative motion is very small. At the natural frequency of the instrument, the relative motion can be very large indeed, and any measurements made near this frequency are bound to be in serious error. Figure 5.5 graphs the relative amplitude versus frequency for the instrument.

Above resonance, a flat uniform output is obtained, proportional to displacement. It is very good practice to keep well away from the resonance condition, for even if correction could be made to the information, the instrument would tend to act as a vibration absorber and would reduce the amplitude to be

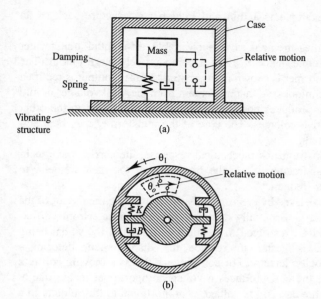

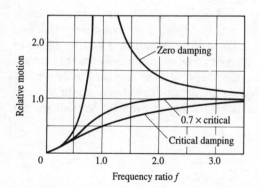

Figure 5.4 *Basic transducer: (a) seismic, (b) torsional.*

Figure 5.5 *Frequency response for seismic displacement transducer.*

measured. This would violate the principle of good measurement which insists that the measuring technique should not interfere with the parameter being measured. It implies that measurement of vibration amplitude should be restricted to frequencies *above* the natural or critical frequency of the instrument. The criterion also applies to velocity sensors, which are essentially the same.

The measuring instrument is heavy, and although it can be shown that the energy absorbed is very small when operated in the correct frequency range, the very fact of attaching the measuring instrument might severely modify the structure. An obvious example is a thin and light beam structure such as an aerofoil.

For displacement measurement, a linear-variable differential transformer (LVDT) is the most suitable, since it gives a comparatively large output. The arrangement of the transformer is shown in Figure 5.6. The AC output is rectified using phase-sensitive techniques, so that movement from the 'null voltage' or mid position in one direction produces positive voltages, and movement in the other direction produces negative voltages. The core of the transformer may be a significant mass in the system.

Regardless of the instrument's mechanical response, the carrier wave technique restricts the signal frequency, i.e. the frequency of the measurement, to about 10% of the carrier frequency.

The *velocity* can be derived by electronically differentiating the output of the LVDT, but differentiation is essentially a 'noisy' process and electronic differentiation has to be partial – restricted in frequency – to avoid the signal being swamped by noise. It is easier to replace the displacement detector + differentiator with a velocity detector. For convenience this is a moving coil in a constant magnetic field; the coil produces a voltage proportional to the rate at which the lines of force are cut, so the voltage is proportional to the velocity in a constant magnetic field. Velocity transducers are therefore simpler in construction than displacement transducers, but because they need to have a light suspension, they are fairly delicate and need to be handled carefully. The only consolation is that it's obvious when they've been damaged.

Simplicity of construction and simplicity of signal acquisition have made velocity transducers much more popular than displacement transducers. But over a range of frequencies, the output tends to be restricted to a smaller range of voltage amplitudes. In mechanical systems, vibration amplitudes tend to increase as the frequency is reduced, whereas velocities tend to be restricted in their peak-to-peak magnitude. This helps to reduce the demands made on amplifying and recording equipment, and it reduces the possibility of overload or readings rendered useless.

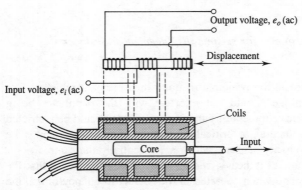

Figure 5.6   *LVDT transducer.*

Accelerometers are used in many vibration monitoring systems. To understand their operation, consider the graph of Figure 5.7; it shows how the relative motion of a seismic transducer is proportional to *acceleration* at frequencies well below resonance down to zero frequency, i.e. in a steady acceleration field such as gravity or in a circular path at constant angular velocity. At these low frequencies, the mass and the vibrating system have exactly the same motion, and acceleration of the mass produces an inertia force proportional to the acceleration.

With damping (about 70% of critical damping), the 'flat' part of the transducer's linear response is pushed up to within 30% of the resonant frequency with the added advantage of reduction in ringing. These curves are easily derived from the equations of motion of a second-order system.

An accelerometer therefore needs to have a resonant frequency *above* the maximum frequency expected in the system under investigation. High natural frequency implies a very stiff system with low mass. This low mass is an advantage for making measurements, and the only restriction is the size of the achievable stiffness. This can be derived from very stiff springs in the cantilever device, but a limit is imposed by the physical size of the strain gauges and by the reduced output produced by smaller beam deflections. These double-cantilever transducers have a voltage output almost inversely proportional to the resonant frequency of the instrument; they are suitable for measurements up to about 70% of the resonant frequency.

By making the transducer spring from a piezoelectric crystal, it is possible to achieve a very high natural frequency; the voltage or charge generated across the crystal can then be used as the force or acceleration-detecting parameter. It is extremely difficult and rarely worthwhile to apply much damping to piezoelectric transducers. The damping is usually less than 5–6% of critical.

Figure 5.7 shows that the linear zone extends to about 10% of resonance. This is not difficult to achieve since piezoelectric accelerometers can be designed with very high natural frequencies – up to 30 kHz is common. Piezoelectric accelerometers would obviously tend to ring or vibrate continuously if an

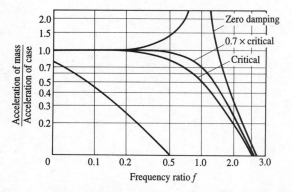

Figure 5.7   *Seismic accelerometer: frequency response to sinusoidal acceleration.*

impulsive acceleration were applied, but in mechanical systems it is difficult to apply such a sudden impulse. The accelerometer 'sees' the applied impulse as a ramp signal, and it can be shown that if the time taken to apply the step or impulse is longer than the time for the transducer to perform about 5–10 vibrations, the ringing will be negligible. Piezoelectric ringing can only be excited using special test equipment such as a ballistic pendulum, essentially a controlled swinghammer which strikes a fixed anvil.

References

American National Standards Institute (1959). *Methods for Calibration of Shock and Vibration Pickups (ANSI S2.2-1959 [R 1976]; ANSI Z24.21-1957 [R 1978]*. New York: ANSI.

British Standards Institution (1988). *Calibration of Vibration and Shock Pick-ups* (BS 6955) London.

Cochin I. and Plass J. R. (1990). *Analysis and Design of Dynamic Systems*. London: HarperCollins.

Doeblin E. O. (1990). *Measurement Systems: Application and Design* 4th edn. New York: McGraw-Hill.

International Organisation for Standardisation (1990). *Standard Methods for the Calibration of Vibration and Shock Pickups (ISO DP5347)*. Geneva: ISO.

6

Vibration signal acquisition

Condition monitoring systems range from integrated standalone units to custom-built rigs comprising separate instruments. But irrespective of the system and its applications, there are certain basics which need always to be borne in mind. No matter how sophisticated the equipment, the quality of its measurements and their interpretation will still depend on the ingenuity of the operator.

The acquisition of a vibration signal is considered in four sections:

- Basic electrical and electronic principles

- Signal types, interference and transmission

- Elementary amplifier and filter design

- Vibration signal acquisition and conditioning

Each of the four topics has generated and continues to generate considerable amounts of literature; recommended texts are cited within each section and collected in a list at the end of the chapter. Other sources are sprinkled parenthetically throughout.

6.1   Basic electrical and electronic principles

The design of electronic instrumentation is a specialised process and is beyond the scope of this text; reference material is readily available (Carr J. J. 1991.

Designers Handbook of Instrumentation and Control Circuits. New York: Academic Press). Several important principles underpin the three stages of primary sensing, variable conversion and electrical interfacing: equivalent circuits, impedance matching, amplifier and filter design and electrical circuit protection during operation (Scott 1987; Pallas-Arney and Webster 1991). Using an example from condition monitoring, we briefly introduce or revise the relevant ideas.

6.1.1 Equivalent circuits

When dealing with equivalent systems and their loading effects, a popular approach is to use a 'systems engineering' method, where all systems – electrical, mechanical, fluidic or any combinations, can be analysed in a similar fashion (Bentley J. P. 1995. *Principles of Measurement Systems*. London: Longman). An elementary procedure is adopted where a simplified equivalent circuit is used to represent all the components of a practical circuit. An important assumption in producing an equivalent circuit is that all the impedances of the components – resistors, capacitors and inductors – have linear characteristics; linear characteristics are also assumed for combinations of these elements and for any semiconductors. Although linear characteristics are not a necessary requirement for producing an equivalent circuit, linear behaviour is both important and desirable in systems where many instruments are connected together.

The general rules for circuit analysis were formulated by Kirchoff and take the form of two laws which apply to circuits comprising impedances and generators, which may be linear or non-linear, constant or variable (Scott 1987).

(1) The total current arriving at a junction of a circuit is zero.

(2) The algebraic sum of the potential differences across components, between any two points in a circuit, is independent of the route taken.

Another useful rule is Thévenin's theorem, which applies to circuits containing two or more sources of e.m.f., active circuits.

An active circuit with a load connected between two terminals, A and B, behaves as if the circuit contained only one source of e.m.f. E having internal resistance r. E is the potential difference between terminals A and B with the load disconnected and r is the resistance of the circuit measured between A and B, with the load disconnected and any sources of e.m.f. replaced by their internal resistances.

As an illustration, consider the example in Figure 6.1. Figure 6.1(a), the circuit model, comprises four distinct features commonly found in instrumentation: (1) voltage source $V(t)$; (2) simple RC filter circuit; (3) voltage amplifier, gain K; and (4) load resistance R. The circuit model may be represented by the block diagram in Figure 6.1(b) and by the equivalent circuit model in Figure 6.1(c).

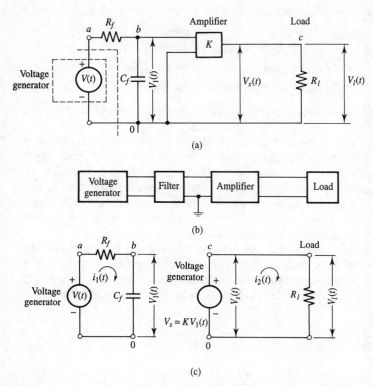

Figure 6.1 *Schematic models of an elementary instrumentation system: (a) circuit, (b) block diagram, (c) equivalent circuit.*

Amplifiers feature widely in all instrumentation work, a topic dealt with in Section 6.3. The amplifiers are normally electronic, and for now it is sufficient to define an amplifier as an instrument or circuit in which a small input signal controls the energy flow from the source to the load. Figure 6.1 shows a voltage isolation amplifier used to isolate a time-varying input voltage $V(t)$ from an elementary load comprising a resistance (capacitors and inductors may also be included). The amplifier in this case is assumed to have no internal resistance, only a constant of proportionality K, which increases the input voltage. Before reaching the amplifier, the signal passes through an elementary filter circuit consisting of a resistor and a capacitor connected as shown (Poularikas and Seely 1990).

Figure 6.1 shows how the original circuit may be reduced to a block diagram then to an equivalent circuit for more convenient analysis. In the block diagram, the transfer function of each part can be found, as a ratio of output variables to input variables, by analysis or by tests; for example, the overall transfer function of the system may be written as

$$V_l(s)/V(s) = KG_f(s)KG_a(s)KG_l(s) \qquad \textbf{(6.1)}$$

where Equation 6.1 is the open-loop transfer function of the complete circuit model and

$$KG_f(s) = \text{transfer function of the filter}$$
$$KG_a(s) = \text{transfer function of the amplifier}$$
$$KG_l(s) = \text{transfer function of the load}$$

The solution of Equation 6.1 depends on the form of each transfer function and the input voltage.

To produce the best combination of circuit components, for minimum energy loss during use, the original system is analysed in the form of an equivalent circuit model. In this case, where the amplifier has no internal resistance, the original circuit can be represented by two separate circuits. The two separate circuits each have a voltage source and the two sources are related via the amplifier gain K. In theory, if it has no internal resistance, the voltage source provided by the amplifier will be unaffected by the load.

The circuits of Figure 6.1(a) and (c) may be analysed separately, but the relationship for the load voltage $V_l(t)$ with respect to a time-varying source voltage $V(t)$ will be the same. Applying Kirchoff's current law to Figure 6.1(a), the current equation at node b after a finite time has elapsed $(t > 0)$

$$i_f(t) - i_c(t) - i_a(t) = 0 \tag{6.2}$$

or in terms of the system component parameters

$$V(t)/R_f - C_f dV_1/dt - i_a(t) = 0 \tag{6.3}$$

Equation 6.3 can be modified to reflect the current relationship for the load and the amplifier gain, this is because

$$i_a(t) = i_c(t) = V_l/R_l \quad \text{or} \quad KV_1(t)/R_l \quad \text{and} \quad V_s(t) = KV_1(t).$$

Therefore the differential equation for the load voltage is

$$\{R_f C_f/K\}dV_l/dt + R_f/R_l V_l = V(t) \tag{6.4}$$

Using a similar analysis over the same time span, the separate circuits in Figure 6.1(c) give the following equations for $t > 0$:

$$\text{Circuit 1:} \quad V(t) = C_f R_f dV_1/dt + V_1(t)$$
$$\text{Circuit 2:} \quad V_l(t) = KV_1(t)$$

and when these equations are combined they produce the same result as Equation 6.4.

Section 5.1.2 has already emphasised the importance of the correct dynamic response of a measurement system, and several solutions can be found for Equation 6.4, depending on how the voltage source varies.

An important result can be found if the source voltage is in the form of a step change over a significant time period. Equation 6.4 can then be solved by separating the variables and integrating over suitable time and velocity limits. With the variables separated, Equation 6.4 becomes

$$\int_0^t dt = \frac{R_f C_f}{K} \int_0^{V_l} \left\{ \frac{1}{V_f - (R_f/R_l)V_l(t)} \right\} dV_l \tag{6.5}$$

and if τ is introduced as a time constant, the following standard integral has to be solved:

$$\int_0^t dt = \tau \int_0^{V_l} \left\{ \frac{dV_l}{(R_l/R_f)V(t) - V_l(t)} \right\} \tag{6.6}$$

where the general solution is

$$t = \tau \log_e \left\{ \frac{(R_l/R_f)V(t)}{(R_l/R_f)V(t) - V_l(t))} \right\}$$

therefore the load voltage V_l is given by

$$V_l(t) = \frac{R_l}{R_f} V(t)\{1 - e^{-t/\tau}\} \tag{6.7}$$

This equation is well known in the dynamics of engineering systems and has been used to describe the dynamic behaviour of many practical systems (Scott 1987).

6.1.2 Impedance matching

Many systems for vibration measurement and analysis connect together several instruments with different electrical impedances. The best representation of the measurand comes from minimising the energy loss (Doeblin 1990, Chapter 2) and this requires impedance matching. Impedance matching has already been introduced in the equivalent circuits of Figure 6.1(c); for when they are connected together, they produce no loss of energy. But this may not be true in practice because all instruments and circuits have their own impedance characteristics.

A sinusoidal voltage produces the following impedance relationships:

- Resistance: $Z(j\omega) = R$
- Capacitance: $Z(j\omega) = 1/j\omega C$
- Inductance: $Z(j\omega) = j\omega L$

When using Laplace transform notation, $j\omega$ is replaced by s.

More complicated circuits produce their own relationships when analysed by Kirchoff's laws and care has to be taken in connecting circuits to produce the minimum loss of power.

The procedure used for efficient impedance matching of circuits and instruments is as follows; the method can be applied to any analogous engineering system (e.g. electrical, mechanical, fluidic, etc.).

Consider the circuit in Figure 6.1. To allow a more accurate representation, the amplifier is assumed to have an internal resistance r ohms; and the load is an elementary voltmeter with load resistance R_l. The modified circuit model is shown in Figure 6.2.

The power equation for individual and complete circuits is important in instrumentation work. In general terms, at the input of each component in a system there exist an input variable q_{i1}, of primary interest, and an associated variable q_{i2}. Their product has the units of power; variation of q_{i1} and q_{i2} identifies the instantaneous change of power. When q_{i1} and q_{i2} are identified as

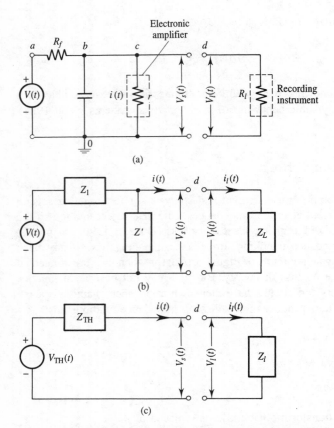

Figure 6.2 *Impedance matching in a measuring system: (a) circuit model, (b) generalised impedance model, (c) equivalent circuit model from Thévenin's theorem.*

physical variables, the generalised input impedance of a component (or system) is given by

$$Z_{\text{input}} = q_{i1}/q_{i2}$$

And the power equation is

$$\text{Power} = (q_{i1})^2/Z_{\text{input}}$$

In the circuit model of Figure 6.2(a), a recording instrument (e.g. voltmeter) is used to measure the unknown voltage $V_s(t)$ at the output of the amplifier. The inclusion of the amplifier internal resistance r means that the value of $V_s(t)$ will not be the same as in Figure 6.1(a). When disconnected from the amplifier and source voltage, the recording instrument has an input impedance

$$Z_1 = V_1(t)/i(t)$$

and a power relationship

$$P_1 = i^2(t)/Z_1$$

To reduce the power loss to a minimum, the input impedance has to be large.

When disconnected from the recording instrument the output voltage from the amplifier will be different from the condition where terminals c and d are joined, and the impedance of the load from the voltmeter has to be included. To simplify the analysis, Thévenin's theorem can be used directly. However, in order to clarify this approach, the full analysis will be shown with reference to Figure 6.2(a) and (b).

Employing the generalised impedance approach (Scott 1987), the circuit model of the measuring system can be represented by the generalised impedance model in Figure 6.2(b). The diagram has been simplified by describing the two components connected in parallel as a single equivalent impedance Z'. All the main impedances are as follows:

$$Z_1 = R_f; \quad 1/Z' = 1/r + sC \quad \text{or} \quad Z' = r/(1 + srC); \quad Z_1 = R_1$$

Apply Kirchoff's laws for the condition when the load is disconnected and the amplifier gain is unity. With a time-varying input voltage $V(t)$, the output voltage from the amplifier is

$$V_s(t) = (Z'/Z_1)V(t) - Z'i(t) \tag{6.8}$$

If the analysis is to be simplified using Thévenin's theorem, the voltage relationship of Equation 6.8 must be equivalent to the equation of the circuit shown in Figure 6.2(c). From Figure 6.2(c), for the same voltage output, the other parameters are

V_{th} = equivalent voltage from Thévenin's theorem

Z_{th} = output impedance of the filter/amplifier circuit from Thévenin's theorem

Z_l = input impedance of the load

For a time-varying voltage at the input, the equation of the amplifier output voltage at the same conditions of current as in Figure 6.2(b) is

$$V_s(t) = V_{th}(t) - Z_{th}i(t) \tag{6.9}$$

Therefore from Equations 6.8 and 6.9, to produce this equivalent system for the same conditions, the following parameters must be used:

$$V_{th}(t) = [r/R_f(1 + srC)] \quad \text{and} \quad Z_{th} = [r/(1 + srC)].$$

By adopting this approach, the system can be directly simplified to the equivalent circuit in Figure 6.2(c). The effects of the load impedance can be found from Equation 6.9 by equating the voltages for the same current conditions at the connections labelled d. This gives the voltage at the load, in this case the voltage indicated by the volmeter, as follows:

$$V_l(t) = (1 + Z_{th}/Z_1)V_s(t) \tag{6.10}$$

Equation 6.10 leads to an important result that applies when circuits and instruments are connected together:

To ensure that true measurements of output parameters are made [in this case a voltage] the input impedance of the load must be high relative to the output impedance of the circuit or system to which the load [voltmeter] is connected.

6.2   Signal types, interference and transmission

A feature of modern measurement systems is their ability to cope with all forms of signal commonly found in industry. Within an industrial environment, signals have to be transmitted efficiently with no significant loss of power. And signals must be protected from the many forms of interference which can pollute their information, producing inaccuracies and making it harder to take the correct decisions. Although some of these features are briefly discussed, the references at the end of the chapter include several texts where all the important information can be found (e.g. Scott 1987; Poularikas and Seely 1990).

6.2.1 Signal types

Three types of signal are normally dealt with in measurement systems:

- periodic signals
- random signals
- discrete signals

Examples of these signals are given in Figure 7.1, but for now it is sufficient to note that periodic signals are the simplest type to deal with. They are continuous in the time domain, and contain either a repeated variation with time of a single frequency or combinations of frequency components. Periodic signals may be described mathematically by a Fourier series and can be analysed by the method called Fourier transformation (Randall 1987). Examples of this process are given in Section 7.4.

A complicated form of signal found in many industries is a random variation in both amplitude and frequency with respect to time; it requires special analytical procedures based on Fourier analysis and statistical properties (Braun 1986). Section 7.5 contains examples of this approach.

With the advent of digital electronics and integrated circuits, modern instrumentation systems handle a discrete form of an originally continuous signal in the time domain. Digital instruments, e.g. measurement transducers and analogue-to-digital converters are now readily available. These instruments produce a discrete form of continuous time domain signals, usually at equal time intervals. Examples of this approach are given in Sections 7.6 and 7.7.

6.2.2 Transmission of signals

An important feature of any instrumentation system is the efficient transmission of data or information. This should be done with the minimum attenuation and degradation of signals. The instruments are not normally located near to one another, so connection from a remote location can be accomplished with a simple coaxial cable, where the inner wire is surrounded by an outer core of insulation and screening. However, cables and transmission lines have their own distributed resistance, inductance and capacitance which may require impedance matching. Often, as a first step, these characteristics are represented by lumped parameters then a simplified analysis is carried out. And instead of voltage amplifiers, high-impedance transducers such as piezoelectric devices use charge amplifiers, therefore it becomes even more important to produce the cable connection with the correct impedance characteristics.

Current sources with suitable cabling have become popular, especially in the process industry, because they can transmit analogue data over large distances. But trends indicate that, in general, for transmission over long distances with a minimum of corruption or attenuation, the signals are converted to digital form. Then ribbon cables are used with 'hard-wired' bus connections on circuit boards. Digital data can be transmitted at very high speed or high baud rate (bits per second). Digital data is covered by various popular standards which deal with transmission, reception and control of data at high speeds:

- ASCII RS-232C, RS-422 and 423 interface standards

- IEEE-488 interface standard (IOtech 1991. *Instrument Communication Handbook*)

Two further methods of data transmission are popular: fibre-optic cables and radio telemetry.

Fibre-optic cable connections are now used extensively in optoelectronic systems. An efficient transmission system requires a controllable light source, fibre-optic cable and a photodecoder (often a silicon photodiode). Fibre-optic cable has several advantages (Doeblin 1990):

- High data transmission rates and low interference over a wide frequency bandwidth.

- Fibre-optic cables carry signals between instruments at high common mode voltage levels, without creating electrical paths between them; this provides high insulation.

- Optical fibres are immune from electromagnetic interference and give good electrical noise rejection.

Many data transmission systems have been developed using visible red light and cables of the polymer fibre type. This allows the efficient transmission of analogue or digital data over short distances, typically 20 m at 10 Mbit s^{-1}.

Cable connections are not always allowable or desirable, e.g. with strain gauges attached to rotating shafts in machinery; data may then be transmitted by FM radio telemetry, irrespective of the system hardware. In general, a telemetry facility comprises two frequency modulated (FM) instruments and an antenna (Doeblin 1990). The transmitter converts time-varying voltages to proportional frequency values in a voltage/frequency converter. The radio signals from the transmitter antenna are received by a second FM instrument or decoder. Very often this instrument consists of inductance-controlled subcarrier oscillators, where a change in inductance causes a proportional change in frequency from a selected centre frequency. The original signal is reconstructed within the decoder to sufficient accuracy for further use.

Each telemetry system requires a specially designed radio frequency antenna having a length of the same order as the wavelength to be transmitted. Recent forms of pulse code modulation system have been used for transmitting data over very long distances, and miniature forms of digital telemetry have been used in rotating machinery strain gauge measurements (Donato V. and Davis S. P. 1973. Radio telemetry for strain measurements in turbines. *Sound & Vibration*, April, **7**, 28–34).

6.2.3 Guarding, shielding and noise

Interference reduces the quality and performance of many circuits. This is mainly due to inadequate guarding or shielding and even contamination by signal noise.

The main sources of interference in circuits are electrostatic fields and magnetic fields. Noise may also be generated by electrical and mechanical components during normal circuit operation; this usually causes random interference. Returning to electrostatic and magnetic fields, a common cause of stray or parasitic capacitance and inductance is the mains supply. Avoid having a difference in potential between earth and ground connections in an instrument; this is because stray or erratic currents may appear in the earth and combine with the resistance of the soil, etc., to form a voltage source – often called a common mode voltage. Due to the asymmetric form of the input circuits of a measuring instrument, the common mode voltage is produced as a differential mode voltage, which introduces an error in the measurements. Problems of this type occur in instruments mounted in a rack, where a current can flow through the rack via the signal ground terminals, whether or not the instruments are connected to ground.

To guard against these important problems, certain practical measures are taken to connect the appropriate parts of a circuit to earth or ground in the correct manner. A good practice is to have a single ground connection for the whole system and to 'earth' all the instruments to that point. This approach will prevent stray currents from appearing and will avoid problems in the instrumentation system. Earthing problems may also be avoided by employing differential input devices. With a differential amplifier the common mode rejection process tends to suppress interference signals.

Another important practical problem concerns the shielding of circuits or instruments. Sensitive electronic parts are placed inside metal cabinets to prevent magnetic fields from entering. The requirements for good shielding may be summarised as follows (Pallas-Arney and Webster 1991).

- The common of the circuits must be attached to ground.

- The common connection must be to the shield at the signal/ground connection only.

- Multiple shields should be connected in series and ultimately to the signal ground and/or common.

- Avoid shield currents at all times.

- A common connection used as a ground must be an equipotential.

It may prove more difficult to shield equipment from magnetic fields, compared with electrostatic fields but there are still some basic rules to follow when setting up:

- Use twisted-pair cable for conductors carrying large currents; wires are a source of magnetic fields.

- Shield circuits with appropriate materials, depending on the applied frequency and magnetic field strength.

- Avoid running wires parallel to a magnetic field; cross magnetic fields at right angles.

- Locate circuits that are sensitive to magnetic fields well away from the source of the magnetic fields and place them at right angles.

- Avoid ground loops and reduce the area of any components that are sensitive to magnetic fields.

- Make any cable lengths as short as possible.

Coaxial cable is a symmetrical conductor so it is generally insensitive to magnetic interference. But wires of any kind may act as either receivers or transmitters of interference.

Finally, we look at random signals; generally called noise, they often reduce the efficiency of a measuring system and must therefore be considered. It may be difficult to pinpoint the reason for noise because it originates from many physical mechanisms. But noise plays a vital role in the handling of signals, especially at very low levels. The noise level may be so large that it swamps the true signal, thus leading to erroneous measurements. Noise can be investigated using the signal-to-noise ratio (SNR); the higher the SNR the better the signal measurement.

6.3   Elementary amplifier and filter design

Many instrumentation circuits have amplifiers and/or filters; certainly this is true of circuits handling analogue voltages. Digital electronics also simulates many analogue amplifiers and filters by suitable components or software subroutines. This section gives general information concerning elementary amplifiers and filters; digital circuits will be covered in Chapter 7.

6.3.1 The operational amplifier

The main building block of an instrumentation amplifier is an operational amplifier (op-amp), nowadays an integrated circuit with two input terminals leading to a very high resistance. Under ideal conditions an op-amp should have an infinite input resistance, zero output resistance and an infinite gain, it is, in fact, a voltage-dependent voltage source with an infinite gain.

Figure 6.3 shows the connections for an operational amplifier used in the differential form; the two inputs are as follows:

- The non-inverting input is termed positive.

- The inverting input is considered negative.

The input voltage is the product of the amplifier gain and the voltage difference at the input:

$$V_O = G(V_{NN} - V_{IN})$$

The output voltage is approximately the same as the power supply voltages $V_{SP}(+)$ and $V_{SN}(-)$; the amplifier saturates at voltages above V_{SP} and V_{SN}. On

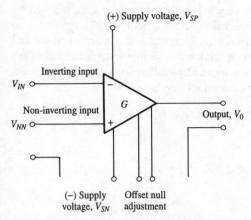

Figure 6.3 *Operational amplifier connections.*

popular commercial amplifiers the output voltage is typically about 1 V (or slightly less) below the power supply voltage of the same polarity.

Basic designs try to match these conditions, but the important practical forms may be studied by considering the op-amp circuit in Figure 6.3. The integrated circuit with high voltage gain G has input and feedback impedances connected as shown. By specifying the form of these impedances, e.g. resistance or capacitance, various types of amplifier may be constructed. Suppose both impedances are resistors ($Z_i = R_i$; $Z_{FB} = R_{FB}$); the very high amplifier gain ($G \rightarrow \infty$) ensures that i_1 and i_2 are approximately zero, thus producing a virtual ground point; the currents in the circuit are

$$i_1 + i_2 = 0$$

or

$$\frac{V_{IN}}{R_i} + \frac{V_O}{R_{FB}} = 0$$

therefore

$$\frac{V_O}{V_{IN}} = -\frac{R_{FB}}{R_i}$$

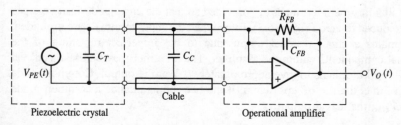

Figure 6.4 *Typical piezoelectric transducer and charge amplifier circuit.*

So, although the open-loop gain G is very high, the closed-loop gain is given by the ratio $-R_{FB}/R_i$, the ratio of the external resistors. This type of amplifier is an inverting and voltage multiplying device. However, if Z_i and Z_{FB} are a resistor and capacitor, respectively, this arrangement produces a simple integrating amplifier; when impedances are reversed, it produces an ideal differentiator. In the first case, the voltage/impedance equation for zero initial conditions is

$$V_{IN}(s)Z_i(s) + V_O(s)Z_{FB}(s) = 0$$

Now when

$$Z_i(s) = \frac{1}{R_i} \quad \text{and} \quad Z_{FB}(s) = sC_{FB}$$

this becomes

$$\frac{V_O(s)}{V_{IN}(s)} = -\frac{Z_i(s)}{Z_{FB}(s)}$$

$$\frac{V_O(s)}{V_{IN}(s)} = -\frac{1}{sR_iC_{FB}}$$

or

$$V_O(t) = -\frac{1}{R_iC_{FB}} \int_0^t V_{IN}\,\mathrm{d}t$$

thus producing an ideal integrating and inverting amplifier. Elementary circuits will not normally operate, so an extra resistor is used for the integrating amplifier; this limits the bandwidth. When the impedances are reversed, using a similar approach, the output voltage equation is

$$V_O(t) = -R_iC_{FB}\frac{\mathrm{d}V_{IN}}{\mathrm{d}t}$$

giving an ideal differentiator.

Before we move on, it is worth noting two important op-amp characteristics:

- input offset voltages

- common mode rejection ratio

Offset null adjustment terminals are provided so that the unwanted offset voltages may be reduced to zero (Figure 6.3). Even when the input terminals are grounded, there remains a small residual output due to the imperfect matching of the individual components within the amplifier. The amplifier may be 'balanced' via the offset null terminals to give approximately 0 V at the output. Common mode rejection limits the use of an op-amp. If the two inputs include a common mode voltage V_{CM}, the actual output is

$$V_O = G(V_{NN} - V_{IN}) + G_{CM}V_{CM}$$

where the effects of the common mode voltage are characterised by the common mode rejection ratio (CMRR) given by

$$\text{CMRR} = 20 \log_{10}\left(\frac{G}{G_{CM}}\right)$$

Notice that the CMRR is expressed in decibels. For good practice, the CMRR should be within the range 60–120 dB and, as a general rule, the CMRR should be as high as possible.

Operational amplifiers are used extensively in modern instrumentation systems. For example, transducers normally produce very small signals which must be amplified before analysis. An instrumentation amplifier can be designed from standard op-amp circuits; the basic design places a voltage follower at each input (Beckwith T. G. 1993. *Mechanical Measurements*. Reading MA: Addison-Wesley). Another common use of op-amp circuits is the charge amplifier, for use with piezoelectric transducers, where the feedback impedance comprises a resistor and capacitor in parallel. With additional circuitry this type of amplifier can be produced to accept inputs from a wide range of piezoelectric devices with different calibration factors (Bruel & Kjaer 1993. *Electronic Instruments*). This makes it easy to set the calibration factor in mechanical units per volt, so the variable can be read directly from a recording device in mechanical units.

A typical circuit for a charge amplifier is shown in Figure 6.4; a transducer and connecting cable with known characteristics are connected to an op-amp. If the transducer, cable and feedback capacitances are C_T, C_C and C_{FB}, respectively, and if the value of the feedback resistor R_{FB} is neglected, the output voltage V_O due to the voltage generated by the piezoelectric device $V_{PE}(t)$ is

$$V_O = \frac{-V_{PE}(t)}{C_{FB} + \left\{\dfrac{C_T + C_C + C_{FB}}{G}\right\}}$$

The second term in the denominator is very small, due to the high open-loop gain G, so the output voltage equation is

$$V_O = -\frac{V_{PE}(t)}{C_{FB}}$$

It transpires that the magnitude of the output voltage for this combined system is independent of the capacitances in the cable and the transducer – an important result.

6.3.2 Elementary filters

A simple network of the form shown in Figure 6.5 may be used as a filter for electrical signals (Lynn 1973). The main elements in the circuit are two impedances Z_A and Z_B; the simplest filters contain just a resistor and a capacitor. As an

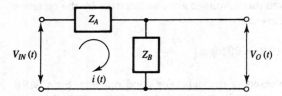

Figure 6.5 *Elementary filter configuration.*

elementary example, assume that Z_A is a resistor R_A and Z_B a capacitor C_B. When a sinusoidal voltage $V_{IN}(t)$ is applied, the currents through the resistor and capacitor are equal:

$$i(t) = \frac{V_{IN}(t) - V_O(t)}{R_A} = C_B \frac{dV_O}{dt}$$

When the input and output voltages have the form $V_{IN}e^{j\omega t}$ and $V_O e^{j\omega t \pm \phi}$, respectively, the general solution of this equation is

$$V_O e^{j\phi} = \frac{V_{IN}}{1 + j\omega R_A C_B}$$

and the phase angle equation is

$$\phi = -\tan^{-1} \omega R_A C_B$$

An amplitude–frequency diagram is drawn to assess the characteristics in the frequency domain for a suitable frequency range. Commonly called a frequency response or Bode diagram, the typical graph for this filter is shown in Figure 6.6; the amplitude ratio V_O/V_{IN} in decibels (dB) is plotted against the logarithm of the frequency ratio ω/ω_c, where ω_c is the cut-off frequency, equal to $1/(R_A C_B)$ or $1/\tau$; $-\tau$ is the time constant of the filter. The Bode diagram clearly shows the characteristics of an elementary RC filter, with a 'flat' response up to a frequency ratio ω/ω_C of approximately unity; the amplitude ratio then drops at a rate of -20 dB per decade. This type of filter is called a low-pass filter. By exchanging

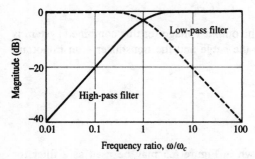

Figure 6.6 *Magnitude versus frequency response: elementary low-pass and high-pass filters.*

the resistor and capacitor in Figure 6.5, the frequency response is made to mirror the low-pass response about the line $\omega/\omega_c = 1$. This configuration may be termed a high-pass filter. A graph of phase versus frequency ratio is of secondary importance for investigating these filter configurations, but it can still be plotted. The phase angle measurement is relatively unimportant for a single channel; and when there are multiple channels, a filter with similar characteristics is used for each channel, thus eliminating the need for phase measurements.

Both analogue and digital filters are in common use for the conditioning and processing of signals. Analogue filters are very often used for the preconditioning of signals, and modern signal analysers have various forms of digital filter for use in signal analysis.

Here is a brief definition of a conditioning processing filter:

A filter is a system which transmits one or more frequency ranges and rejects others; the transmitted ranges are called passbands, the rejected ranges stopbands.

We now give a brief description of analogue filters; digital filters will be described in Section 7.6. Analogue filters are produced in two basic forms:

Passive filters comprise resistors, capacitors and inductors connected in an appropriate fashion, usually on a circuit board.

Active filters are constructed from op-amps (Clayton G. B. 1979. *Operational Amplifiers*. London: Butterworth).

Active and passive filters can be constructed to produce the following categories. They are classified by the frequency ranges transmitted or rejected and illustrated in Figure 6.7.

Low-pass filters operate in the low-frequency regions.

High-pass filters operate in the high-frequency regions.

Bandpass and bandstop filters discriminate for or against particular frequency bands.

A modern design approach is to manipulate a system's characteristic equation to produce a particular frequency response or shape. Many important filter frequency response shapes are now available, such as Butterworth, Bessel and Chebychev – low-pass filters that are frequently used in signal acquisition and processing (Burr-Brown, Tucson AZ).

6.3.3 Application of amplifiers to transducers

As a general guide, LVDTs (linear-variable differential transformers), usually have a sufficiently high output from the demodulator that they can be fed directly into oscilloscopes, frequency analysers, etc. The same normally applies to moving-coil and variable-reluctance transducers. Few problems are generally

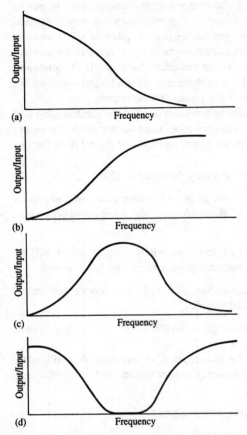

Figure 6.7 *Approximate frequency response of filters: (a) low-pass, (b) high-pass, (c) bandpass, (d) bandstop.*

encountered in matching them to other instruments, since the characteristic impedance is low and common mode interference plus mainsborne noise rarely reach troublesome levels (Dally J. W. and Riley W. F. *et al* 1993. *Instrumentation for Engineering Measurements*. New York: Wiley, Chapter 6).

Strain gauge transducers, in the form of double-cantilever accelerometers, require special strain gauge amplifiers. The main characteristics of such amplifiers are long-term stability and freedom from common mode interference, achieved using balanced amplifiers at the input stages, the CMRR can be 90–100 dB, so the SNR should be about 100 000 to 1. These amplifiers are expensive and must not be overloaded, even though overload protection diodes, etc., are usually fitted. Common mode rejection is assisted by proper application of some simple rules. Most important is the elimination of ground loops, or circulating currents within connecting leads. Adequate screening from ambient magnetic fields is also desirable.

Quartz-crystal amplifiers are usually designed to amplify the *charge* on the crystal rather than the voltage appearing across it (Sydenham P. H. and Hancock N. H. 1992. *Introduction to Measurement Science and Engineering*. New York: Wiley, Chapter 5). This has several important advantages:

- Signals can be measured accurately at low, medium and high frequencies. This is because the time constants of the system depend on the selection of the amplifier feedback resistors, even though the op-amp input impedance is extremely high.

- Cables of more than 100 m can be used without adversely affecting performance; the length of the input cable is theoretically unimportant.

- Common mode problems are virtually eliminated because the amplified quantity is charge, not terminal voltage.

6.4   Vibration signal acquisition and conditioning

Successful measurements require high-quality data. This means that every effort is made to quantify the errors involved before acquisition and conditioning. This section highlights the main problems that lead to errors in vibration signal acquisition and conditioning.

6.4.1 Signal acquisition

The first and vital step in vibration monitoring is to capture a recording of vibration levels, normally as time passes. The degree of difficulty in producing a vibration recording increases with the complexity of the instrumentation (Figure 5.3). One option involves only transducers, conditioning amplifiers and a tape recorder (Racal) or oscilloscope. If they are chosen to suit the application (e.g. transducer characteristics) and the impedance matching is correct, it is possible to make a time domain recording of vibration levels. The recording may be inspected in the normal way by playing back the tape and noting areas for further viewing or analysis.

Two other options require microprocessor-based instruments. These instruments, the real-time analyser and dedicated computer, are designed to use recording and analysis techniques which must be clearly understood before applying them. In all three options, the instrument impedance matching must be correct before acquiring any recordings. Continuous time recordings are taken in the first option (normally called analogue signals). But the first stage in the other two options, for both real-time analyser and dedicated computer, is to produce discrete time recordings (digital signals), to do this, each device employs an analogue-to-digital converter, and it must be clearly understood that this produces an approximate form of the original continuous signal. The sampling of a continuous signal and subsequent storage in its discrete form will not work unless some important conditions are met:

- The type of signal must be known, whether periodic, transient or random, so some preliminary work needs to be done with an original continuous recording.

- The analogue-to-digital conversion requires a data sampling rate. The sampling rate is generally selected from Shannon's sampling theorem, which ensures that fictitious discrete data is not captured, generally known as aliased data. Anti-aliasing filters are also used, either analogue or digital. Digital filters are produced by a software subroutine in an instrument (Wowk 1991).

- Following analogue-to-digital conversion, the data has to be stored in blocks for later use. Each instrument must have sufficient memory and appropriate software so the blocks of discrete data can be manipulated and calculations carried out.

6.4.2 Avoiding aliasing

Aliasing is overcome by adopting a rule called *Shannon's sampling theorem* (Braun 1986): a continuous signal which contains no significant frequency components above f_c Hz may in principle be recovered from its sampled version, if the sampling interval is less than $1/2f_c$ seconds. For a sampling interval of h seconds, the sampling rate is therefore $1/h$ samples per second. To comply with the theorem, 2 samples per cycle are required to define a frequency component in the original data. The highest frequency which can be defined by sampling at a rate of $1/h$ samples per second is therefore $1/2h$. Apart from selecting the correct sampling frequency, before sampling it is normal to filter the data with a low-pass filter of cut-off frequency f_c. The digitising of a vibration recording is shown in Figure 6.8(a). An aliased signal is illustrated in Figure 6.8(b), where a combined signal contains frequencies at 1 Hz and 10 Hz. When the signal is shown as separate components, several points on the two frequency components coincide. If the signal were sampled at these points it would be impossible to tell which frequency was being considered; this would give spurious and incorrect results.

6.4.3 Analogue-to-digital conversion

Analogue-to-digital conversion, also called digitising, is an important part of signal acquisition. It is generally carried out by sample-and-hold devices under the control of a microprocessor; Figure 6.9 shows the important steps. The continuous signal measured by a transducer system is first filtered by a low-pass

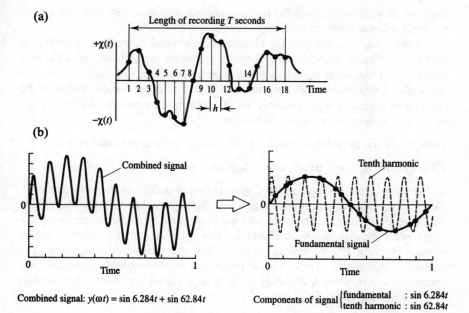

Figure 6.8 *(a) Digitising a vibration recording. (b) Misinterpretation of signals – aliasing.*

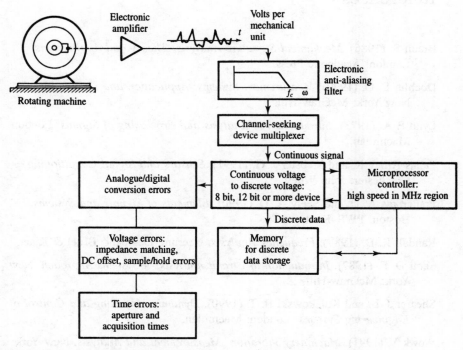

Figure 6.9 *Analogue/digital conversion: control and errors.*

filter, to avoid aliasing, then fed to the analogue-to-digital converter via a multiplexing (channel-seeking) circuit.

Thereafter, depending on the bit representation and operational features, the continuous signal in its discrete form is stored at equal time intervals in a memory. All the control of this process is normally carried out by a microprocessor. A vital feature of this procedure is that an approximate version of the continuous signal is now available in the instrument memory. But errors are associated with the conversion procedure.

Voltage errors: impedance matching, DC offset, sample-and-hold errors.

Time errors: aperture errors, acquisition time errors.

A wide range of analogue-to-digital conversion devices are now available; indeed many companies produce intelligent digital signal processors (Amplicon, Brighton, East Sussex). These devices are controlled by a microcomputer (PC). Some computing power is available on this type of signal processor, which releases the host PC for management of files and interactive computing. The onboard intelligence of the signal processor ensures that the critical time operations are controlled, allowing work to be done with real-time data or data from a computer file.

References

Braun S. (1986). *Mechanical Signature Analysis: Theory and Applications*. London: Academic Press.

Doeblin E. O. (1990). *Measurement Systems: Application and Design* 4th edn. New York: McGraw-Hill.

Lynn P. A. (1973). *Introduction to Analysis and Processing of Signals*. London: Macmillan.

Pallas-Arney R. and Webster J. G. (1991). *Sensors and Signal Conditioning*. Chichester: John Wiley.

Poularikas A. D. and Seely A. D. (1990). *Elements of Signals and Systems*. Boston: PWS–Kent.

Randall R. B. (1987). *Frequency Analysis*. Naerum, Denmark: Bruel & Kjaer.

Scott D. E. (1987). *Introduction to Circuit Analysis: A Systems Approach*. New York: McGraw-Hill.

Shearer J. L. and Kulakowski B. T. (1990). *Dynamic Modelling and Control of Engineering Systems*. London: Macmillan.

Wowk V. (1991). *Machinery Vibration: Measurement and Analysis*. New York: McGraw-Hill.

7

Vibration signal analysis

Analysis of a vibration signal may vary from the elementary – direct measurement of amplitude and period for periodic vibration with a constant amplitude in the time domain – to the sophisticated application of several mathematical methods to the analysis of a random variation in both the time and frequency domains. Any form of analysis aims to provide information on the amplitude of vibration and the predominant frequencies of vibration. The type of analysis and its degree of complexity depends on the type of vibration.

Research and development over the past 20 years produced many new methods for signal analysis, well documented up to the mid 1980s by Braun (1986). The basic tools for signal analysis; systems analysis and signal processing are described in sufficient detail to give a good background of the subject. Literature is available on applications of these techniques and on new methods of signal processing but for now the aim is to highlight some useful applications.

Consider the block diagram in Figure 7.1. The typical stages of vibration signal acquisition and analysis are shown for rotating machinery vibration. Vibration signals are conditioned then acquired for assessment before being passed on to the signal analyser; the information may now be subdivided into three groups:

- amplitude information
- correlation information
- frequency information

Analysis methods for condition monitoring have been developed largely within these three groups, so each will be considered in turn, some applications identified and literature cited; the references are collected in a list at the end of the chapter. For rotating machines it may be assumed that the majority of

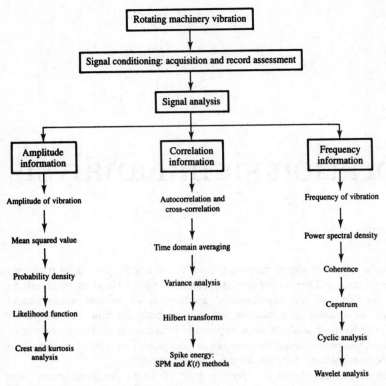

Figure 7.1 *Typical stages of vibration signal acquisition and analysis.*

vibration signals are stationary and ergodic (certainly for part of their time history) and so we proceed on that basis, at least to begin with.

Initially, in any analysis, important information may be provided simply by displaying and measuring the amplitude of a periodic vibration, particularly when a single frequency is involved. More complicated signals (e.g. multifrequency periodic, random or transient) require electronic instruments and or software packages (Digis software package) to provide measurements of the mean square value and probability density function of a variable (Newland 1993).

Consider the block diagrams in Figure 7.2. To find the mean square value of a signal, in the time domain, the original signal will be manipulated by a squaring and averaging process. The result is the mean square value of the signal contained in a recording of length T seconds. Thus the equation of this process for the mean square value of a variable $x(t)$ is

$$\bar{x}^2(t) = \frac{1}{T}\int_0^T x^2(t)\mathrm{d}t$$

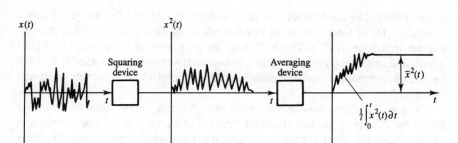

Figure 7.2 *Finding the overall mean square value of a random variable in the time domain.*

An extension to the method of finding the probability density function is to apply a method which operates on acquired data to find the maximum likelihood function (Cempel 1991). Indeed this approach assumes that the data has some statistical property, then the likelihood function will be an estimator. Crest and kurtosis analysis have been applied in engineering systems where impact effect are common; a good example is the estimation of damage to rolling element bearings (Mellor D. J. 1990. Element Bearing Damage Detection. Wallingford: Condition Monitoring Ltd).

Correlation information is found from well-established theory which allows both the autocorrelation and cross-correlation functions to be found from data (Bendat and Piersol 1993). Early applications of this work are found in the literature on vehicles (Dodds 1975), and electronic instruments are available for measuring and displaying these functions (Bruel & Kjaer 1995. *Frequency and Signal Analysers*). Within the past 10 years, extensions to the methods involving time domain signals have been published, e.g. time domain averaging by suitable filters (Ljung 1985) and the application of both variance and Hilbert transform analysis to systems which produce repetitive data (Braun 1986). And several procedures have been adopted for roller bearing diagnostics, where the shock pulse method and spike energy (*IRD Mechanalysis, Columbus OH*) method have been employed (Strum 1991) for extracting useful information from time signals. A new method to estimate rolling element bearing damage is currently being applied (Hills 1994), where a time-dependent factor is calculated from the product of peak acceleration and RMS acceleration; the factor is quoted as the ratio of the acceleration product at installation to the acceleration product at the time of investigation.

A substantial amount of the signal analysis literature deals with producing frequency information from a time series; as this is so important for condition monitoring, it is vital to understand the theoretical background. The frequency of a simple vibration can be calculated from its period. But the vibration signals from most rotating machinery contain harmonics of the fundamental rotation frequency; so the data has to be analysed by Fourier methods. Fourier analysis is well established for periodic and random vibrations, both continuous and discrete

(Lynn 1986). The continuous Fourier transform (CFT) and the discrete Fourier transform (DFT) have been used to great effect via an algorithm called the fast Fourier transform (FFT). The FFT may be programmed on a suitable digital mainframe or desktop computer and it is now relatively straightforward to find the frequency components of vibration signals on proprietary spectrum analysers (Section 7.2).

Figure 7.3 shows how to obtain the overall mean square value of a variable $x(t)$ in the frequency domain (Doeblin 1990). The signal is first inspected using a narrow bandpass filter (notch type) and the frequency content over the filter bandwidth is measured before being passed on to squaring and averaging devices, which produce the mean square value of the signal in the frequency domain. If the data has many frequencies of importance, this process has to be applied several times to find the frequency content or power spectral density. Modern spectrum analysers or comparable software must therefore be capable of performing the stages simultaneously, so that results can be produced in real time (Section 7.7). This approach generally produces a power spectral density of the form

$$S(\omega) = \frac{\pi \bar{x}^2(\omega)}{\Delta \omega}$$

Although power spectral density methods have been applied to general signal analysis problems, including condition monitoring, several extensions to the method have had to be developed to tackle important problems in machinery: input/output relationships, transient vibration and non-stationary variables. For example, the coherence function allows the input and output functions of a system to be related (Bendat and Piersol 1986). Another important extension to spectral analysis is cepstrum analysis, successfully applied to the assessment of vibration damage in gearboxes (Muralidharan 1994) and to the assessment of production levels and quality in paper making (Brown 1991). Recent publications report the application of cyclic analysis for non-stationary vibration monitoring of reciprocating machines (Serridge 1991).

Signal analysis encompasses many complicated procedures, so as a guide for future study, the remaining sections contain basic information on typical vibration

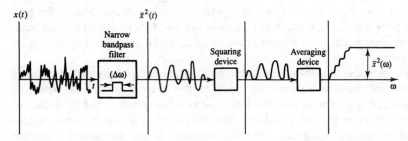

Figure 7.3   *Finding the overall mean square value of a random variable in the frequency domain.*

signals, modern analysis instruments, physical interpretation of important para-meters, discrete Fourier analysis and the fast Fourier transform, digital filters and the use of windows in real-time analysis.

7.1 Typical vibration signals

Figure 7.4 shows five typical signals from rotating machinery recorded in the time domain:

- Periodic signal with constant amplitude and frequency.
- Transient signal with periodic, single frequency.
- Periodic signal with added frequency components (harmonics).
- Random vibration signal: both amplitude and frequency are random.
- Discrete signal obtained at a constant time interval or sampling rate.

The discrete signal occurs after digitisation of a continuous time signal.

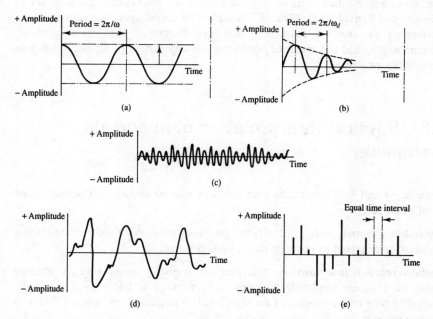

Figure 7.4 *Typical vibration signals in rotating machinery: (a) periodic waveform, constant amplitude, single frequency; (b) transient waveform, periodic; (c) periodic signal with added frequency components; (d) random waveform; (e) discrete signal.*

7.2 Modern instruments for signal analysis

Modern instruments employ averaging methods to produce amplitude information, e.g. the mean square value of a signal over a time period. And the statistical properties of a vibration signal may be estimated and presented as probability densities or autocorrelation functions, (Newland 1993) interpreting the amplitude information in ways that are useful to engineers. However, care must be taken when using these approximate methods because only a few have probability distributions amenable to mathematical solutions (e.g. the Gaussian or normal distribution); this is especially true in the area of random signal analysis.

Frequency is an important parameter in vibration analysis; in systems with multiple degrees of freedom – most practical machinery – a large number of vibration frequencies may have to be found. For an elementary signal, where only one frequency is present, a measurement of the period of the vibration signal in the time domain is sufficient to calculate the frequency of vibration. A more extensive procedure is required for more complicated time variations, which have many periodic signals added together, and for some random signals. Fourier methods generally provide adequate analysis, and certain stringent conditions are applied where the form of the signal demands them. Fourier analysis obtains the fundamental and its harmonics for periodic and random signals. It can also be applied to discrete data using an algorithm called the fast Fourier transform (FFT) (Bendat and Piersol 1993). The FFT has allowed digital spectrum analysis to be performed on modern frequency analysers (some of them developed by Schlumberger) and did much to popularise the application of signature analysis on rotating machines.

7.3 Physical interpretation of important parameters

Figures 7.4 and 7.5 illustrate the parameters of interest and given below are some brief definitions.

Probability density in its crudest form specifies a measure of the probability that an event is expected to occur within a certain range.

Autocorrelation functions are functions which give a measure of the average value of a signal amplitude variation. The average is taken over a specific recording time of the product of the signal and its original form, time-shifted by a known amount.

Spectral (or power spectral) density is a parameter which allows the frequency content of a signal to be specified; the Fourier components or the frequency spectrum are used for a periodic signal. But the spectral density of a random

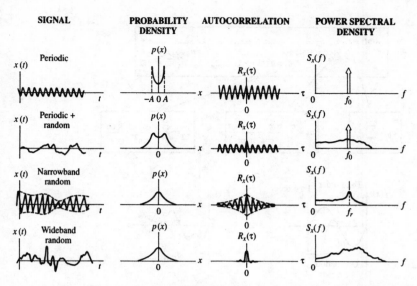

SIGNAL	PROBABILITY DENSITY	AUTOCORRELATION	POWER SPECTRAL DENSITY

Figure 7.5 *Typical vibration signals and important theoretical functions.*

signal is given by the Fourier transform of the autocorrelation function, often specified as a mean square spectral density function.

It is common to find these three parameters on modern measuring instruments. Spectral density is very well used in machinery health monitoring and thus warrants special attention. Two important areas need to be considered:

- Fourier analysis of continuous and discrete recordings
- Fourier analysis windows

We will give the theoretical methods a physical interpretation; this should make them easier to understand and may help with the identification of spurious results, not unusual in practical situations.

7.4 Fourier analysis

Figure 7.6 shows how a periodic vibration signal may be described by a Fourier series. The form of the series indicates that the waveform consists of several waves which have their own constant or Fourier coefficient multiplied by a sine or cosine term. This is done from 0 to the harmonic of order N. To facilitate analysis, an approximation to the Fourier coefficient may be used over one period of vibration $0-2\pi$ radians.

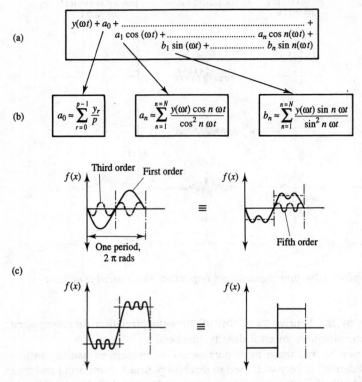

(a)

$y(\omega t) + a_0 + \dots\dots\dots\dots\dots\dots\dots\dots\dots\dots\dots\dots\dots\dots +$
$\qquad a_1 \cos(\omega t) + \dots\dots\dots\dots\dots\dots\dots\dots a_n \cos n(\omega t) +$
$\qquad b_1 \sin(\omega t) + \dots\dots\dots\dots\dots b_n \sin n(\omega t)$

(b)

$$a_0 \approx \sum_{r=0}^{p-1} \frac{y_r}{p} \qquad a_n \approx \sum_{n=1}^{n=N} \frac{y(\omega t)\cos n\,\omega t}{\cos^2 n\,\omega t} \qquad b_n \approx \sum_{n=1}^{n=N} \frac{y(\omega t)\sin n\,\omega t}{\sin^2 n\,\omega t}$$

(c)

Figure 7.6 *Fourier analysis in engineering: (a) Fourier series for periodic waveform from 0 to 2π radians; (b) engineering approximations for Fourier coefficients; (c) typical application of a sine wave plus harmonic components to approximate a square wave.*

If over one period of vibration a waveform is divided into 12 or more equally spaced amplitude ordinates (y_0 to y_p), the approximations of the Fourier coefficients are

$$a_0 \approx \sum_{r=0}^{p-1} \frac{y_r}{p}$$

$$a_n \approx \sum_{n=1}^{n=N} \frac{y(\omega t)\cos n\omega t}{\cos^2 n\omega t}$$

$$b_n \approx \sum_{n=1}^{n=N} \frac{y(\omega t)\sin n\omega t}{\sin^2 n\omega t}$$

The application of Fourier methods to vibration analysis may be seen from the construction of a square wave signal by adding harmonics to a sine wave (Figure 7.6); approximately 80 components are needed to achieve an acceptable square wave. Frequency analysis aims to find the harmonic components that are required to approximate a recorded vibration signal. Using the discrete Fourier transform and the FFT algorithm, it has been extended from continuous periodic signals to random signals and discrete signals, the subject of the next section.

7.5   Discrete Fourier analysis and the fast Fourier transform

Figure 7.7 shows how a sequence of r discrete data points may be truncated to a sequence of N data points for discrete Fourier analysis. An efficient algorithm for the procedure of Figure 7.7 is the fast Fourier transform (FFT). In general terms, the approach demands that the full sequence of discrete points $\{x_r\}$ is partitioned into a number of shorter sequences, e.g.

$$y_r = x_{2r}; \qquad z_r = x_{2r+1}$$

$$r = 0, 1, 2 \ldots \left(\frac{N}{2} - 1\right)$$

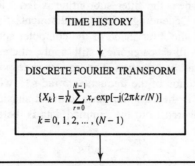

TIME HISTORY

DISCRETE FOURIER TRANSFORM

$$\{X_k\} = \frac{1}{N}\sum_{r=0}^{N-1} x_r \exp[-\mathrm{j}(2\pi k r/N)]$$

$$k = 0, 1, 2, \ldots, (N-1)$$

FAST FOURIER TRANSFORM ALGORITHM

1. Partition $\{x_r\}$ into shorter sequences $\{x_{2r}\}$ and $\{x_{2r+1}\}$ for $r = 0, 1, 2, \ldots, (\frac{1}{2}N - 1)$

2. Take DFT of short sequences:

$$\{Y_k\} = \frac{1}{N/2}\sum_{r=0}^{(\frac{1}{2}N-1)} y_r \exp\{-\mathrm{j}[2\pi k r/(N/2)]\}$$

$$\{Z_k\} = \frac{1}{N/2}\sum_{r=0}^{(\frac{1}{2}N-1)} z_r \exp\{-\mathrm{j}[2\pi k r/(N/2)]\}$$

3. Combine in form. $\{X_k\} = \frac{1}{2}\left[Y_k \exp\{-\mathrm{j}[2\pi k r/(N/2)]\}\right]$ for $k = 0, 1, 2, \ldots, (\frac{1}{2}N - 1)$

Figure 7.7   *Use of the FFT method in discrete signal analysis.*

Compare this with the direct approach:

- N multiplications

$$\{x_r\} \times \exp(-2\pi j k r)/N$$

have to be performed to obtain each of the N values of X_k – the new finite sequence formed from $\{x_r\}$.

- N^2 operations are needed to calculate the full sequence $\{X_k\}$. The FFT approach reduces the work to a number of operations of order $N \log_2 N$. The FFT partitions the original discrete sequence $\{x_r\}$ into a number of shorter sequences; it calculates the DFT of each shorter sequence then combines them to yield $\{X_k\}$, the full discrete Fourier transform of $\{x_r\}$.

7.6   Digital filters

Digital filters play an important part in discrete signal analysis, especially when accurate spectral densities are required. Digital or sampled data filters have a completely different construction to analogue filters. A digital filter is created using a software program, where the filter subroutine is fed with the sampled version of the input signal. Depending on the requirements for analysis, the sampled input is modified by the filter shape of the computer subroutine.

Although the analogue filter categories still apply, there is a different interpretation of bandwidth and its selection. The frequency response in sampled data systems is a periodic function of the frequency ω rad s^{-1}, which is repeated every $2\pi/h$ rad s^{-1}, where h is the sampling time to avoid aliasing. The highest frequency which can be recovered from the sampled data is therefore

$$\hat{\omega} = \frac{2\pi}{2h} = \frac{\pi}{h} \text{ rad s}^{-1}$$

So it is convenient to classify a digital filter according to its effect on frequency components in the frequency range

$$-\frac{\pi}{h} < \omega < \frac{\pi}{h} \text{ rad s}^{-1}$$

This is the maximum frequency range occupied by an adequately sampled input signal.

Many standard filters have been constructed for sampled data analysis. Indeed, digital filtering of a signal is equivalent to applying a discrete Fourier transform, or to filtering a signal with N parallel filters, modulus H_K which are centred around the frequencies kdf. In this context df is the frequency spacing of the DFT. If the filter centre frequency does not coincide with that being analysed, there will be leakage over all the frequency range. This problem may be alleviated by applying window functions.

7.7 Windows for real-time analysis

Real-time analysers (such as the 3595 series from Solartron, Farnborough) provide frequency analysis of the whole vibration signal in all frequency bands of interest on the basis of the digital samples captured by the instrument and within the frequency bandwidth selected. The windowing function must therefore be chosen with care. Some windows required to analyse the most common types of signals are shown in Figure 7.8; other windows may be available on modern instruments. Efficient analysis of periodic and stationary random signals generally calls for the Hanning window; transient signals, pseudorandom signals and other special signals are usually analysed with the rectangular window because it gives the best estimates of spectral density.

Apart from any errors inherent in the choice of windowing function, several other errors often arise at this stage:

- maximum amplitude errors

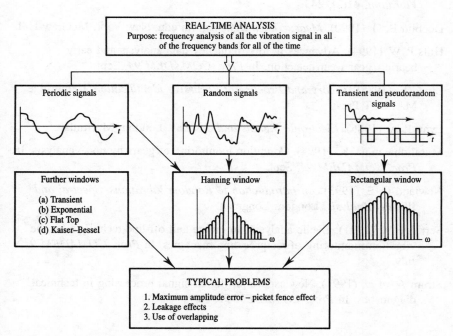

Figure 7.8   *Use of windows in real-time analysis.*

- leakage effects of filters
- overlapping of data during analysis

Correction factors are normally applied to alleviate them.

References

Bendat J. S. and Piersol A. G. (1986). *Random Data: Analysis and Measurement Procedures* 2nd edn. New York: Wiley.

Bendat J. S. and Piersol A. G. (1993). *Engineering Applications of Correlation and Spectral Analysis* 2nd edn. Chichester: John Wiley.

Braun S. (1986). *Mechanical Signature Analysis: Theory and Applications*. London: Academic Press.

Brown D. N. (1991). Envelope and cepstrum analysis: two vibration techniques helping to maintain production output and paper quality. *TAPPI J.*, **74**, 5–8.

Cempel C. (1991). *Vibroacoustic Condition Monitoring*. Chichester: Ellis Horwood.

Dodds C. J. (1975). Partial coherence in multivariate random processes. *Sound & Vibration*, **42**(2), 243.

Doeblin E. O. (1990). *Measurement Systems*, 4th edn. New York: McGraw-Hill.

Hills P. W. (1994). Advances in low frequency data analysis and early bearing/gear wear detection. In *Proc. COMADEM'94*, Sept.

Ljung L. (1985). *Theory and Practice of Recursive Identification*. Cambridge MA: MIT Press.

Lynn P. A. (1986). *Electronic Signals and Systems*. London: Macmillan.

Muralidharan M. K. (1994). Condition monitoring of gears by noise analysis. In *Proc. COMADEM'94*, Sept.

Newland D. E. (1993). *An Introduction to Random Vibrations, Spectral and Wavelet Analysis*. London: Longman.

Serridge M. (1991). Cyclic analysis: an online and off-line technique for the vibration monitoring of reciprocating machines. In *Proc. COMADEM'91*, July.

Strum A. *et al.* (1991). New aspects of digital signal processing in technical diagnostics. In *Proc. COMADEM'91*, July.

Rotational and reciprocating balance

8.1 Rotation of flexible shafts on flexible supports

8.1.1 General points

To remedy rotational unbalance, it is essential to understand the principal features of mass/elastic behaviour in rotating flexible shafts:

- Even shafts of complex shape are comparatively easy to model.

- Bearings are difficult to model; they may have different stiffnesses in the vertical and horizontal directions, causing parametric vibrations so that one or more of the system parameters is a function of time, e.g. effective stiffness, mass or damping.

- Damping in the bearings is difficult to determine and is still the subject of research.

8.1.2 Classes of problem

Problems may be classified into four categories:

- Very stiff systems in which the maximum running speed is well below the

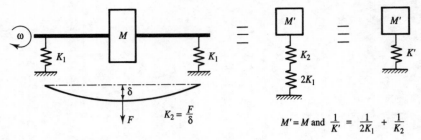

Figure 8.1 *A flexible shaft on flexible supports.*

lowest natural frequency, i.e. $\omega \ll \sqrt{K'}/M'$ for the simple case shown in Figure 8.1.

- A rigid rotor ($K_2 = \infty$) on flexible supports ($K_1 \neq \infty$).

- A flexible rotor ($K_2 \neq \infty$) on rigid supports ($K_1 = \infty$).

- A flexible rotor ($K_2 \neq \infty$) on flexible supports ($K_1 \neq \infty$).

8.1.3 Flexible shafts and rigid shafts

If, in the course of calculating the natural frequencies of a system, the stiffness of a particular element is set to infinity and the natural frequencies as calculated agree with measured values, to within the accepted tolerance, it is justified to assume the element is rigid.

Does an easy way exist for determining whether or not an element is rigid? Simple cases can be discovered by inspection, but the more complex cases rely on various rules of thumb built up over the years. For example, if the dynamic bearing deflection is less than or equal to one-tenth of the dynamic shaft deflection, the bearing is rigid. A rigid shaft has also been defined as one whose lowest pinned/pinned natural frequency is 50% higher than the running speed (Morton P. G. 1986. Private communication).

A flexible element is an element whose stiffness has to be quantified. If its stiffness were taken as infinity, any results would be unacceptably crude, but they might serve as first-order approximations or as starting values in more sophisticated solutions.

The first of the four classes in Section 8.1.2 is a special case; problems only arise when the out-of-balance is very large. It is much more common for natural frequencies to have a magnitude similar to the running speed, but not equal or trouble would result. There is often at least one natural frequency below the normal running speed, so the system must pass through resonance during run-up and rundown. The design and operation of rotating systems depend upon which of the three remaining classes it belongs to.

More detail will be given later about different modal forms of vibration (whirling) of flexible shafts and how the out-of-balance forcing the vibration depends strongly on the deflected form of the whirling shaft, which in turn depends on the rotational speed of the shaft (i.e. where the out-of-balance is speed dependent).

Out-of-balance is one cause of vibration, another cause is the in-service change that all machines undergo. Out-of-balance can be caused by erosion, corrosion, wear, thermal distortion, build-up of deposits, component failure, bolts becoming loose, etc., etc. All of this can be considered as wear and tear, easily remedied by adequate inspection and maintenance procedures.

Lastly, operators make alterations to machines to change performance or output; in so doing they may inadvertently alter the state of balance and/or the natural frequencies of the system. Vibration levels sometimes increase because a natural frequency has been shifted near to the running speed, one of its multiples or some other frequency of excitation. The increase in vibration has even been caused simply by an increase in the unbalance of rotating parts.

8.1.4 An unbalanced rotor

An unbalanced rotor is defined by its final state of unbalance, which must be less than a certain limit or tolerance (ISO 1940–1973, BS 5265: Part 1: 1979). Perfect balance is unattainable, although sometimes a rotor may be so well balanced that the final state of unbalance cannot be detected by the measuring equipment. However, it would be unwise and incorrect to define this as perfect balance, since the actual state of unbalance is unknown. The correct conclusion is that the residual unbalance is less than the minimum quantity discernible by the measurement method.

8.1.5 Rotational balance of a rotor

There are two common methods to quantify rotational balance, which is expressed by stating the residual unbalance.

RMS vibrational velocity

The RMS value of the vibrational velocity of the main bearings at the running speed is measured and quoted. This is called the vibration severity and is most suited to machines with constant speed. If there are several running speeds, the vibration severity should be quoted for each one. The amplitude of the sinusoidal component of the velocity at the running speed is sometimes quoted as an alternative.

Physical unbalance of the assembly

The physical unbalance of the rotating assembly in g mm can be quoted with angular positions in each of two planes; this is only suited to rigid rotors. Machinery users sometimes assume that, if the unbalance expressed in g mm is inside the tolerance, the machinery behaviour will be safe and acceptable at all speeds of rotation. But this may not be true; to be completely safe, the dynamic response of the rotor must be investigated when it is installed in the assembly – that includes the supports and any coupled machinery.

8.1.6 Variable-speed machinery and resonance

Jet engines, car engines and other machinery may run at any speed within an operating range, and they must be designed to do this safely. If resonant speeds exist within the operating range, ways must be found to avoid the machinery dwelling there. A resonance may have to be passed through while making a speed change; if so, this passage should be rapid enough to prevent vibrations with dangerously large amplitudes. If dwelling at resonant speeds is unavoidable, it may be necessary to add damping to limit the amplitude of vibrations. Bear in mind that components, such as compressor blades, can also experience resonant vibrations, excited by pulsations usually at frequencies which are multiples of the running speed.

8.1.7 Vibration isolation

Strictly speaking, the state of unbalance in all rotating systems is a function of the rotational speed. The only practical exception is a system with one degree of freedom operated at speeds well above its unique natural frequency. Systems with more than one natural frequency can exhibit similar behaviour when operated at speeds well in excess of the highest natural frequency of significance and well below the next natural frequency. It is a standard design strategy to opt for systems with a single degree of freedom and to run them only at speeds which are greater than the unique natural frequency, a strategy utilised by the designers of the De Laval steam turbine and also in the design of the domestic spin-drier. This has been called vibration isolation because, providing that the speed is not allowed to drop below a certain minimum value (about twice the natural frequency), it can be varied at will by the operator in the knowledge that the (quite small) amplitude of the vibrations will remain substantially constant. The centrifugal force due to out-of-balance is balanced internally, leaving only a small residual force to be transmitted to the support, so the building housing the machinery is unaffected by speed changes. The building is effectively isolated from the machinery, and the machinery is isolated from the building.

Introduction of damping into the elastic support reduces both the amplitude of vibration and the force transmitted to a foundation at resonance. But at rotational speeds well above resonance, although the amplitude of vibration is much the same as in the case of zero damping, the foundation force is greater and increases continuously with increasing speed; this is because the damping force is proportional to the vibrational velocity. The design of supports having damping is a compromise and consists of arranging the values of stiffness, mass and damping coefficient so that the graph of foundation force versus frequency has an acceptable shape.

8.1.8 Car wheel balancing

Because it is done to very crude limits, the balancing of a car wheel (and tyre) is relatively cheap. A poor final state of unbalance can be tolerated due to the high degree of damping in the system, i.e. the shock absorber dissipates the energy of the vibrations in the form of heat. And genuinely user-friendly equipment is provided so that the measurements can be made and the correction weights attached by semi-skilled personnel. The equipment for balancing car wheels is special-purpose equipment in which the instrumentation and signal processing is chosen to suit the individual rigid rotor configuration. Normal balancing machines are designed to accommodate rotors in many shapes and sizes, and to balance all of them to close limits.

8.1.9 Balancing a flexible turbo-alternator

It is expensive to balance a turbo-alternator set because the system is so complex and difficult to model accurately (see section 8.1.1). Among other things, its dynamic behaviour depends on the shaft temperature distribution and bearing oil pressure, so the balancing operation needs the expertise of an engineer who has specialised in rotor dynamics and instrumentation.

8.1.10 Rotational balance and vibration problems

Rotational balance can cure some vibration problems, but not all of them. It will be a frustrating, expensive and an unsuccessful enterprise to attempt a rotational remedy if the force causing the vibration is generated by one or more of the following:

- reciprocating out-of-balance bodies
- unbalanced electrical forces
- geometrical misalignment, perhaps of couplings

- bearing seals that rub
- loose parts moving continuously or intermittently

Attempting to balance a system at a resonant speed is also a pointless exercise, and a flexible shaft balanced at one speed may be badly unbalanced at another.

With any vibration problem, the first thing to do is determine its nature; and that's where modern spectrum analysers prove so valuable. Any resonances must be detuned, loose parts must be made secure, and so on. Only when one is satisfied that out-of-balance centrifugal forces are the cause of unacceptably high vibration at the shaft speed, only then can one attempt to balance the shaft rotationally.

8.1.11 Balancing a flexible rotor

A flexible rotor can be balanced for one speed only. For example, consider a long flexible rotor with a known unbalanced mass m at its centre, as shown in Figure 8.2(a). This unbalance can be corrected by the addition of two masses m_1, as shown in Figure 8.2(b). Now when the rotor is run at a speed of just over one-half of its first critical speed, as in Figure 8.2(c), the dynamic deflection increases the

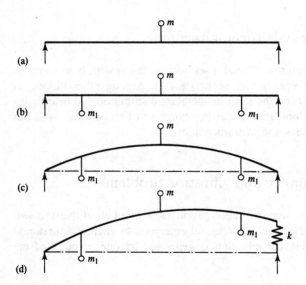

Figure 8.2   *A flexible rotor can be balanced for one speed only: (a) unbalanced flexible rotor; (b) correction using two masses* m₁; *(c) deflection when the rotor runs at just over ½ times its first critical speed; right-hand bearing support is given stiffness* k.

unbalance and decreases the corrective effect of the masses m_1. Therefore let the right-hand bearing support have stiffness k, as in Figure 8.2(d). With or without bearing flexibility, such a system must be balanced at its running speed because the state of unbalance is speed dependent. The deformed shape whirls as the shaft rotates at the *critical whirling speed*.

8.2 Critical whirling speed

8.2.1 A light shaft carrying a centrally mounted disc

Consider a centrally mounted disc of mass M on a light shaft running at a speed of ω rad s^{-1} between two bearings as shown in Figure 8.3(a). Let the centre of gravity G of the disc be at a radial distance e (eccentricity) from the disc centre S. S will describe a circle and the shaft will whirl. The centrifugal force is a vector rotating at ω rad s^{-1}; it can be resolved into vertical and horizontal components giving two harmonically varying forces. The whirling motion can be considered as simultaneous vibrations equal in amplitude, in the vertical and horizontal planes caused by the components of the centrifugal force.

Let B be the point of intersection of the plane AA and the straight line connecting the bearing centres, as shown in Figure 8.3(a). BSG is drawn in a straight line because it is the only configuration that gives equilibrium, as shown in Figure 8.3(b). A centrifugal force of $M\omega^2(x+e)$ acts outwards, tending to increase x. An elastic force of Kx acts inwards, tending to reduce x, i.e. it tends to straighten the shaft. For a steady whirling motion, the elastic and centrifugal forces are in equilibrium:

$$Kx = M\omega^2(x+e) = M\omega^2x + M\omega^2e$$

Solving for the shaft deflection gives

$$x = e\frac{\omega^2}{(K/M) - \omega^2} = e\frac{\omega^2}{\omega_n^2 - \omega^2}$$

$$x = e\frac{1}{(\omega_n^2/\omega^2) - 1}$$

so

$$x \to 0 \quad \text{as} \quad \omega \to 0$$

$$x \to -e \quad \text{as} \quad \omega \to \infty$$

$$x \to \infty \quad \text{as} \quad \omega \to \omega_n$$

This relation is plotted in Figure 8.3(c) and we can draw the following conclusions:

- For $\omega < \omega_n$, G lies outside S.

- For $\omega \gg \omega_n$, G lies inside S and is 'at rest on the axis of rotation'.

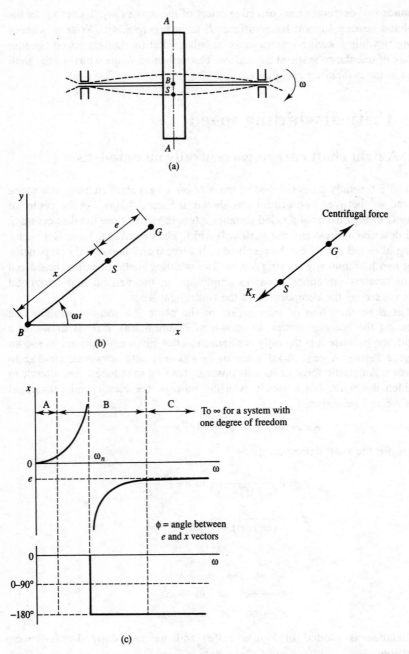

Figure 8.3 *(a) A disc on a flexible rotating shaft. (b) Shaft and bearing centres in relation to* G. *(c) Amplitude and phase versus speed.*

- For a perfectly balanced shaft $(e = 0)$ at resonance $Kx = M\omega^2 x$. Since both forces are proportional to x, the state of equilibrium is 'indifferent'. For a system with no damping, a small sideways perturbation causes an infinite growth of amplitude.

- For an unbalanced shaft $(e = a$ finite value$)$ at resonance, $x = \pm\infty$ – a mathematical discontinuity.

- In frequency ranges A and C small changes in ω make no (significant) difference to the amplitude or phase.

- Ranges A or C must be used for the vibration measurements in rotational balancing calculations.

In range B a speed difference of even a few r.p.m. produces big changes to amplitude and phase. Rotational balancing operations require that the rotor be run, brought to rest and run again with trial masses attached. Experience shows that several runs are often needed. To attempt balancing operations at a speed close to a resonant speed would require that, for every run, the speed would have to be brought back to the same speed to within very close limits indeed – not practically attainable. Well away from a resonance, small differences in the speeds of balancing runs make no significant difference to measurements of amplitude and phase. The 'soft bearing' balancing machine employs vibration isolation so that all measurement runs are well into range C. The rotor is run up to speed with the bearings locked, i.e. on high stiffness. When the rotor is up to speed, the bearings are unlocked and become very soft, allowing the rotor to run at well above its first natural frequency.

8.3 Balancing a rigid rotor

8.3.1 Static and dynamic balance

When the rotor in Figure 8.4 is mounted on knife-edges, its unbalance m_u can apparently be corrected by adding m_c in the position shown. This is known as static balance. When the rotor is actually spun in bearings, there exists a centrifugal couple whose plane of action is the plane containing m_u and m_c, as shown in Figure 8.4(a); this plane rotates at the same speed as the rotor. The rotating couple can be eliminated if *two* correction masses m_1 and m_2 are used to give static and dynamic balance, as shown in Figure 8.4(b). Two correction planes are needed to achieve rotational balance.

8.3.2 How to produce a general state of unbalance

Consider a perfect rotor unbalanced by concentrated masses $m_1, m_2, m_3, \ldots, m_n$ at radii $r_1, r_2, r_3, \ldots, r_n$ respectively; the mass distribution of all conceivable real rotors can be created in this way. Figure 8.5 shows m_n and its unbalanced

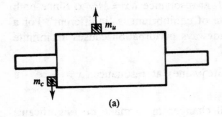

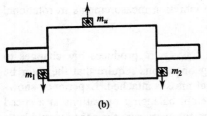

Figure 8.4 *(a) The unbalance m_u of the stationary rotor is corrected by mass m_c. This is static balance; the configuration produces a centrifugal couple when the rotor is spun. (b) The centrifugal couple is eliminated by replacing m_c with two masses m_1 and m_2. The rotor now has dynamic balance.*

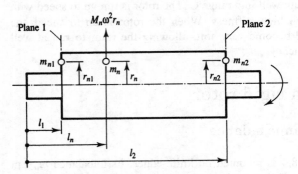

Figure 8.5 *Two masses having the same effect as one.*

centrifugal force $m_n\omega^2 r_n$. The effect of m_n on the equilibrium of the rotor can be achieved by removing m_n and replacing it with two masses, m_{n1} at r_{n1} in plane 1 and m_{n2} at r_{n2} in plane 2. The centrifugal effects of m_{n1} and m_{n2} must meet the same equilibrium conditions as the centrifugal effect of m_n on its own:

$$m_{n1}\omega^2 r_{n1} + m_{n2}\omega^2 r_{n2} = m_n\omega^2 r_n \quad \text{(same force effect)}$$

$$m_{n1}\omega^2 r_{n1}l_1 + m_{n2}\omega^2 r_{n2}l_2 = m_n\omega^2 r_n l_n \quad \text{(same moment effect)}$$

The two unknowns are the mr products $m_{n1}r_{n1}$ and $m_{n2}r_{n2}$; they could be determined if m_n, r_n and l_n were known (ω^2 cancels out).

Repeating the process for masses 1, 2, 3, etc., and by vector summation, a single mass is produced in each plane, as shown in Figure 8.6(a) and (b).

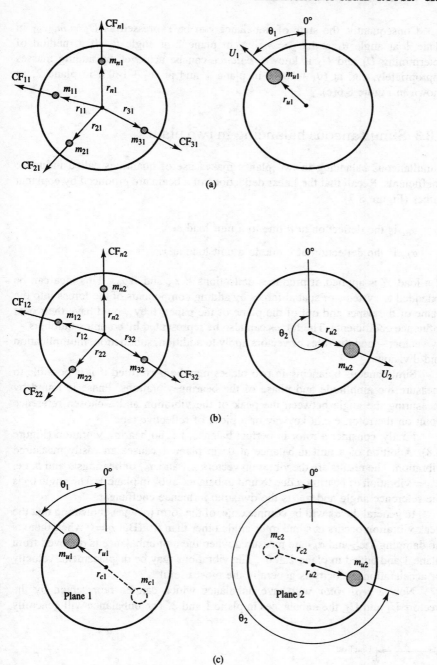

Figure 8.6 *(a) Plane 1: one mass has the effect of many. (b) Plane 2: one mass has the effect of many. (c) Positioning of correction weights.*

Consequently, the state of unbalance can be expressed as $U_1 = m_{u1}r_{u1}$ in plane 1 at angle θ_1 and $U_2 = m_{u2}r_{u2}$ in plane 2 at angle θ_2. If a method of determining U_1 and U_2 is known, *balance* can be achieved by adding masses appropriately, e.g. at $(\theta_1 + 180°)$ in plane 1 and at $(\theta_2 + 180°)$ in plane 2, as shown in Figure 8.6(c).

8.3.3 Simultaneous balancing in two planes

Simultaneous balancing in two planes makes use of quantities called influence coefficients. Recall that the linear deflections of a beam are produced by coplanar forces (Figure 8.7):

- α_{ax} is the deflection at a due to a unit load at x.

- α_{bx} is the deflection at b due to a unit load at x.

If a load W is applied, it produces deflections $W\alpha_{ax}$ and $W\alpha_{bx}$. This idea can be extended to systems of spatial forces by adding components of the forces into the plane of the paper and out of the plane of the paper; they would have their own influence coefficients. The forces can also be represented by complex numbers – by vectors – and the rules of vectors apply to addition, subtraction, multiplication and division.

Simultaneous balancing in two planes may be performed if it is possible to measure the amplitude and phase of the bearing vibrations. Phase is found by measuring the angle between the peak of the vibration and a chosen reference point on the rotor, e.g. a keyway or a piece of reflective tape.

Firstly, consider a rotor in perfect balance, i.e. no bearing vibration (Figure 8.8). Addition of a unit unbalance at $0°$ in plane 1 causes an easily measured vibration. The results are the vibration vectors α_{a1} and α_{b1} of bearings a and b, i.e. α_{a1} = vibration of bearing a due to unit unbalance at $0°$ in plane 1. The angle $0°$ is the reference angle and α_{a1} is the dynamic influence coefficient.

In general the αs will be complex, i.e. of the form $(a + jb)$, indicating that the peak vibration occurs at some rotor angle other than $0°$. (But $b = 0$ when there is no damping.) α_{a2} and α_{b2} are measured when the unit unbalance is removed from plane 1 and added in plane 2 at $0°$. The vibrations may be displacement, velocity or acceleration; velocity is generally the most useful.

Now a *real rotor* will have unbalance which can be represented by the vectors U_1 and U_2, the unbalances in plane 1 and 2. The unbalance will generally

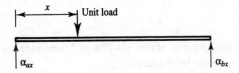

Figure 8.7 *Influence coefficients for a point load on a beam.*

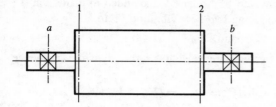

Figure 8.8 *Dynamic influence coefficients: consider a perfect rotor experiencing unit unbalance in plane 1 then in plane 2.*

occur at some angle other than the $0°$, U_1 and U_2 will be complex numbers. The vibration vectors for bearings a and b can be written as follows:

$$V_a = \alpha_{a1} \wedge U_1 + \alpha_{a2} \wedge U_2 \quad \text{therefore} \quad U_1 = \frac{\alpha_{b2}}{\Delta} \wedge V_a - \frac{\alpha_{a2}}{\Delta} \wedge V_b \quad (8.1)$$

$$V_b = \alpha_{b1} \wedge U_1 + \alpha_{b2} \wedge U_2 \quad \text{therefore} \quad U_2 = \frac{\alpha_{a1}}{\Delta} \wedge V_b - \frac{\alpha_{b1}}{\Delta} \wedge V_a \quad (8.2)$$

where $\Delta = \alpha_{a1} \wedge \alpha_{b2} - \alpha_{b1} \wedge \alpha_{a2}$

Before any trial weights are added, the rotor is spun at speed ω and the vibration vectors representing the vibration of bearings a and b are measured (Figure 8.9) as

$$V_a \quad \text{and} \quad V_b$$

If a unit trial unbalance weight W is added in plane 1 at $0°$ and the rotor is again spun at ω, the vibration vectors for planes 1 and 2 are now

$$V_{a1} \quad \text{and} \quad V_{b1}$$

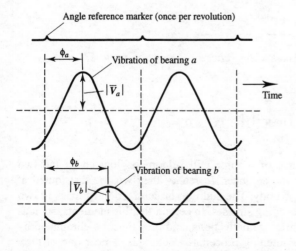

Figure 8.9 *Vibration traces due to natural rotor unbalance.*

The unit trial unbalance W is removed from plane 1 and added to plane 2 at $0°$; the rotor is again spun at ω, producing two new vibration vectors:

$$V_{a2} \quad \text{and} \quad V_{b2}$$

Figure 8.10 shows how the vectors are related to the dynamic influence coefficients:

$$
\begin{aligned}
V_a + W\alpha_{a1} &= V_{a1} & \text{therefore} \quad \alpha_{a1} &= (V_{a1} - V_a)/W \\
V_b + W\alpha_{b1} &= V_{b1} & \text{therefore} \quad \alpha_{b1} &= (V_{b1} - V_b)/W \\
V_a + W\alpha_{a2} &= V_{a2} & \text{therefore} \quad \alpha_{a2} &= (V_{a2} - V_a)/W \\
V_b + W\alpha_{b2} &= V_{b2} & \text{therefore} \quad \alpha_{b2} &= (V_{b2} - V_b)/W
\end{aligned}
\tag{8.3}
$$

Equations 8.1 and 8.2 allow us to determine the unbalances U_1 and U_2 in planes 1 and 2. A computer program would usually be written or some commercial software purchased. Two methods are available to achieve balance:

- Attach a correction mass having the same *mr* products as U_1 and U_2 in planes 1 and 2 *opposite* to U_1 and U_2 to cancel them out.

- Remove material at the *same* angles as U_1 and U_2 in planes 1 and 2 by drilling and/or grinding away material.

Material removal involves a greater element of trial and error than correction masses.

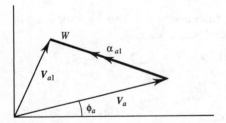

Figure 8.10   *Influence coefficients are related to vibration vectors.*

8.4   Balancing a flexible rotor

The masses causing unbalance of a rotor will seldom be located in the two correction planes chosen by the engineer. In reality they will be distributed at unknown points throughout its body. The addition or subtraction of mass in two planes will produce internal bending stresses. In certain flexible elements, such as long cylinders, this is likely to cause deflections, and they must be eliminated by an additional centrifugal straightening procedure. Rotors at or near one of their critical speeds are subject to this form of elastic bending. A Cardan shaft may operate at 70% of its fundamental critical speed, possibly higher, because it is dynamically balanced by three weights: one placed in the middle plane and the other two in the endplanes. Rotors in power station generators operating at speeds of 1500–3000 r.p.m. – near the first or even the second harmonic of their flexural critical speed – require additional compensation planes in which weights can be added. (Figure 8.11)

Although at least *three* compensation planes are required to provide accurate correction of rotors spinning at speeds near their *fundamental* flexural critical speed, *four* planes (Figure 8.11) are needed to prevent deflection of rotors operating near the *first harmonic* of their flexural critical speed, and *five* (Figure 8.12) are required for rotors running near the *second* harmonic of their flexural critical speed.

In order to perform correct balancing in accordance with the rotor requirements, some balancing machines are designed to measure unbalances in several planes without attaching any additional equipment to the rotor under test. For direct measurement of the deflections of long cylinders, described above, an additional vibration pick-up may be mounted in the middle plane (Figure 8.12). But remember that balancing flexible rotors is a specialist job, usually performed by experts.

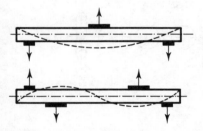

Figure 8.11   *Correction weights for first and second whirling modes.*

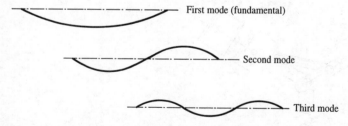

Figure 8.12   *The first three natural whirling modes for a rotor simply supported at its ends.*

8.5   Multicylinder engine unbalance

The majority of modern engines have multiple cylinders, and from the analysis in Section 4.4, reciprocating effects will produce another source of vibration which may affect the condition of machinery. A popular cylinder arrangement in cars and small commercial vehicles is to have four in-line cylinders; the V-bank arrangement is common in heavy vehicles, railway locomotives and shipboard engines.

Examples of both arrangements are shown in Figures 8.13 and 8.14. The in-line engine depicted has four cylinders in the vertical plane; the cylinders are connected to the crankshaft with the cranks arranged in the sequence 1 and 4 then 2 and 3 with respect to top dead centre (TDC). A six-cylinder V-bank arrangement is also shown with a plan view of the crankshaft and crank sequence.

To provide balancing of all the reciprocating effects in these engines, the analysis of the single reciprocator has to be extended to deal with inertia force effects and inertia couple effects. The following brief analysis for in-line and V-bank engines illustrates the important features using specific examples.

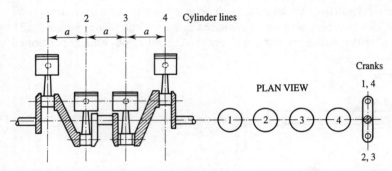

Figure 8.13 *Four-cylinder in-line engine.*

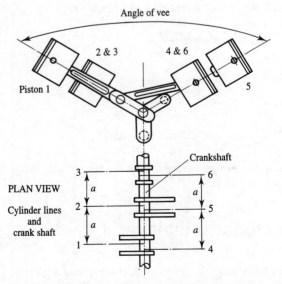

Figure 8.14 *Six-cylinder V-bank engine.*

8.5.1 Regular in-line engines

Consider the in-line engine arrangement of Figure 8.15; the cylinder, pistons, connecting-rods and cranks all have the same dimensions. Two other features are also important:

- The firing order or crank sequence on approaching top dead centre (TDC) is designed to give a good crankshaft arrangement; this applies to both petrol and diesel engines. A good arrangement assists with the partial balancing of reciprocating forces and couples; it also avoids large fluctuations in the engine torque, which helps to avoid exciting critical speeds in the engine speed range.

- The spacing between each cylinder centreline (normally equal) is important in the analysis of the effects due to inertia couples.

The firing order of the engine in Figure 8.15 is

$$1 \quad 2 \quad 4 \quad 6 \quad 5 \quad 3$$

This is the order in which each crank and piston approaches TDC. The crank spacing angle is δ, measured for each crank with respect to crank 1, positioned at TDC. A reference line is drawn at the centreline of the in-line arrangement, so that the inertia couple effects can be analysed, and the distances to each cylinder centreline are marked off as follows:

$$a_1 = a_6, \quad a_2 = a_5 \quad \text{and} \quad a_3 = a_4$$

In each pair the magnitudes are the same but the signs are opposite. All the inertia force effects due to the reciprocating masses (at the pistons) will be in the same line, and only rotational masses at the cranks will produce out-of-line forces at the

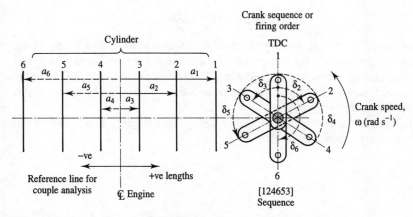

Figure 8.15 *In-line engine: cylinder lines and crank sequence.*

engine speed. Using Equation 4.12, the nth order force at each piston, applied to the frame of the engine and transmitted via the crankshaft and bearings, will be

$$F = C_n \cos n(\theta + \delta) \qquad (8.4)$$

And the nth order couple transmitted to the frame will be

$$C = a[C_n \cos n(\theta + \delta)] \qquad (8.5)$$

Now convert Equations 8.4 and 8.5 to give the total force and couple effects:

$$\text{Total } n\text{th order frame force } (FF_n) = \sum C_n \cos n(\theta + \delta) \qquad (8.6)$$

and

$$\text{Total } n\text{th order frame couple } (FC_n) = \sum aC_n \cos n(\theta + \delta) \qquad (8.7)$$

Rewrite in complex number notation:

$$FF_n = \frac{C_n}{2}\left[e^{jn\theta} \sum e^{jn\delta} + e^{-jn\theta} \sum e^{-jn\delta} \right] \qquad (8.8)$$

and

$$FC_n = \frac{C_n}{2}\left[e^{jn\theta} \sum ae^{jn\delta} + e^{-jn\theta} \sum ae^{-jn\delta} \right] \qquad (8.9)$$

When solving these equations for a particular engine, the first important variables to consider are n, δ and a. For the case where $n = 1, 2, 4$ and 6 and $\delta = 60°$, the forward- and reverse-rotating vectors $\sum e^{jn\delta}$ and $\sum e^{-jn\delta}$ are zero for the first- second- and fourth-order conditions; the sixth-order terms give the total value of the force.

So the form of the sixth-order frame force is

$$FF_6 = C_6[6e^{j6\theta} + 6e^{-j6\theta}]$$

or

$$FF_6 = 6C_6 \cos 6\theta \quad \text{newtons}$$

This identifies a vertical force with a simple harmonic form and acting at six times the engine speed. The magnitude of the force will be governed by the size of the sixth-order component C_6, and a small value of C_6 will hardly influence the engine components. But in the absence of first-order forces, no balance-weights need be attached to the cranks.

To analyse the effects of the inertia couple, a unit length is assumed between each cylinder centreline, such that

$$a_1 = 2.5 \qquad a_6 = -2.5$$
$$a_2 = 1.5 \qquad a_5 = -1.5$$
$$a_3 = 0.5 \qquad a_4 = -0.5$$

From Equation 8.9, the first-order frame couple has the form

$$FC_1 = \frac{C_1}{2}\left[e^{j\theta}\sum ae^{j\delta} + e^{-j\theta}\sum ae^{-j\delta}\right] \text{N m}$$

And from the information given, the forward- and reverse-rotating vectors $\sum ae^{j\delta}$ and $\sum ae^{-j\delta}$ have the following values:

$$\sum ae^{j\delta} \approx 7.21\, e^{-j(14°)}$$

and

$$\sum ae^{-j\delta} \approx 7.21\, e^{j(14°)}$$

giving a total first-order couple of

$$FC_1 = 7.21\, C_1 \cos(\theta - 14°)\ \text{N m}$$

This equation describes a couple phased at 14° behind the angular position of crank 1. To reduce this effect, balance-weights are required outside the lines of cylinders 1 and 6 at the angular phasing dictated by the position of the inertial couple vector; the magnitude is calculated with respect to the distances between each balance-weight.

8.5.2 Vee twin-bank engines

Equation 4.13 is a general expression for the reciprocating force in cylinders inclined at an angle α to the vertical axis. This equation can be used to describe the reciprocating force in a vee twin-cylinder arrangement where the cylinders are inclined at $\pm\alpha$ degrees to the vertical axis. Then the total force transmitted to the frame due to the reciprocating effects, in line with the vertical axis, is

$$FF_n = C_n \cos_n(\theta + \alpha)e^{-j\alpha} + C_n \cos_n(\theta - \alpha)e^{+j\alpha} \quad \textbf{(8.10)}$$

All the cylinders, pistons, connecting-rods and cranks have the same dimensions, so C_n is the same for each reciprocator. With this in mind, and using complex number notation, the total force transmitted to the frame is

$$FF_n = \frac{C_n}{2}\left[e^{jn\theta}2\cos(n-1)\alpha + e^{-jn\theta}2\cos(n+1)\alpha\right] \quad \textbf{(8.11)}$$

The analysis of the inertia couple transmitted to the frame for several vee twin arrangements, attached to the same crankshaft, is similar to the analysis for the in-line regular engine in Equations 8.5 and 8.7. It is now possible to calculate the nth order frame couple for a variety of vee twin arrangements, as shown in Figure 8.14. If the centrelines are distance a apart, with crank sequence angle δ, the nth order frame couple is

$$FC_n = \frac{C_n}{2}\left[e^{jn\theta}\sum ae^{jn\delta}2\cos(n-1)\alpha + e^{-jn\theta}\sum ae^{jn\delta}2\cos(n+1)\alpha\right] \quad \textbf{(8.12)}$$

Then the effects of this couple can be found for a particular vee twin engine with known values of angles α and δ, distance a and the harmonic order number. Proceed in a similar fashion to the in-line example of Section 8.5.1.

References

Den Hartog J. P. (1985). *Mechanical Vibrations*. New York: Dover.

Hong S. W. and Lee C. W. (1989). Identification of bearing dynamic coefficients by using measured and computed unbalance responses in flexible rotor-bearing systems. *Proc. IMechE. Pt. C*, **203**, 93–101.

Morrison J. M. L. and Crossland B. (1970). *An Introduction to the Mechanics of Machines*. London: Longman.

Muszynska A. (1986). Modal testing of rotor bearing systems. *Int. J. Modal Analysis*, July, 15–34.

9

Bearing and gear vibration

All rotor malfunctions and dynamic failures produce a great deal of energy, dissipated from the system through bearings and their supports. The bearing locations are good places to obtain an early warning of impending failure, so it is important to understand the mechanism of bearing vibrations.

In a typical gearbox, vibration levels can arise from several sources. Even when the transducers are mounted on a bearing housing, vibration effects can still be observed – caused by other bearings, remote from the sensor, as well as other gear meshing frequencies. Indeed, extraneous signals may swamp any vibrations from the bearing under measurement. It is therefore essential to isolate specific bearing frequencies, particularly in rotating element bearings, to enable the faulty bearing or related component to be identified.

9.1 Elementary theory for bearing frequency analysis

This elementary theory is taken from a private communication of M. R. T. Thomas (1992): Elementary Theory for Bearing Frequency Analysis.

Figure 9.1 illustrates the geometry of a typical bearing, and the relevant parameters:

n = number of rolling elements
d = diameter of rolling element
D = cage diameter (average of inner and outer diameters)

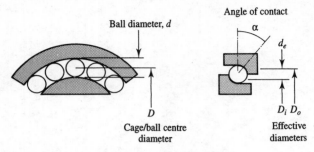

Figure 9.1 *Typical rolling element bearing.*

f = rotational speed (of inner relative to outer race)
α = element contact angle

9.1.1 Basic geometrical relationships

The relationships between diameters are

$$d_e = d \cos \alpha$$

$$d_i = D - d_e$$

$$= D[1 - (d/D) \cos \alpha]$$

$$d_o = D + d_e$$

$$= D[1 + (d/D) \cos \alpha]$$

And the relationships between the speeds are

$$Df_c = -f_b d \qquad \qquad \textbf{(9.1)}$$

$$\frac{1}{2}D_i f = -f_b d \qquad \qquad \textbf{(9.2)}$$

where

f_c = cage frequency
f_b = ball spin frequency
f = angular velocity of inner relative to outer

The angular velocity of the outer race is normally zero, i.e. $f_o = 0$.

9.1.2 Cage rotational frequency

The cage rotational speed relative to the inner shaft rotational speed can be determined from the geometry of the bearings. Equations 9.1 and 9.2 give

$$Df_c = \frac{1}{2}D_i f$$

$$f_c = \frac{1}{2}\frac{D_i}{D}f$$

But from the geometry

$$D_i = D\left(1 - \frac{d}{D}\cos\alpha\right)$$

$$\text{therefore} \quad f_c = \frac{1}{2}f\left(1 - \frac{d}{D}\cos\alpha\right) \tag{9.3}$$

9.1.3 Inner race defect frequency

Defects on the inner race depend on the rate of impact of ball-bearings against a point on its surface. The rate of impact will depend on the number of ball-bearings and on the velocity of the centres of the ball-bearings (equivalent to the cage frequency) relative to the inner race.

$$f_{inner} = n(f - f_c)$$

Substituting for f_c from Equation 9.3 gives

$$f_{inner} = n\{f - \tfrac{1}{2}f[1 - (d/D)\cos\alpha]\}$$

which becomes

$$f_{inner} = \frac{nf}{2}\left[1 + (d/D)\cos\alpha\right]$$

9.1.4 Outer race defect frequency

Defects on the outer race use a similar argument.

$$f_{outer} = n(f_o - f_c)$$

But the outer race is fixed, so

$$f_o = 0 \text{ and}$$

$$f_{outer} = -nf_c$$

substituting for f_c from Equation 9.3 gives

$$f_{outer} = \frac{-nf}{2}\left[1 - (d/D)\cos\alpha\right]$$

9.1.5 Ball defect frequency

A defect on a ball (rolling element) will impact on the inner and outer races, and will depend on the rotational speed of the ball relative to the cage. The actual frequency will be twice the ball rotational frequency relative to the cage rotational frequency.

$$f_{ball} = 2(f_b - f_c)$$

Substituting for f_b from Equation 9.1

$$f_{ball} = -2[(D/d)f_c + f_c]$$
$$= -2f_c[(D/d) + 1]$$
$$= -2f_c(D/d)[1 + (d/D)]$$

Substituting for f_c from Equation 9.3, and replacing d in the brackets with $d \cos \alpha$, gives

$$f_{ball} = -f(D/d)[1 - (d/D)\cos\alpha][1 + (d/D)\cos\alpha]$$
$$\therefore \quad f_{ball} = -f(D/d)\{1 - [(d/D)\cos\alpha]^2\}$$

Note the replacement of d to simplify the equation; the sign indicates rotation of the ball relative to the inner race.

9.2   A practical technique for bearing frequency analysis

In addition to the formulae developed in Section 9.1, another technique has proved successful for finding the predominant bearing frequencies of practical systems during motion. It is based on the use of velocity diagrams for geartrains and is generally known as the epicyclic gear analogy (Lim K. P. 1992. The detection of machinery malfunction. *PhD Thesis*, University of Strathclyde).

9.2.1 Epicyclic gear analogy

In epicyclic gearing it is essential to establish the speed ratios where the vibrations of selected components within the geartrain are controlled at a precise level. A simple and reliable method for bearing analysis assumes that no slip occurs between any of the rolling elements of a good bearing. The simplicity of the analysis lies in the use of an elementary graphical technique, where diagrams do not need to be drawn to scale, and in the fact that each of the rotating elements rotates in either a clockwise direction or an anticlockwise direction. Thus the relative motion of each rolling element can be drawn on a horizontal line, with clockwise rotation to the right of the stationary point, or pole, and anticlockwise rotation to the left of the stationary point. The initial analysis is carried out with the cage of the bearing held stationary and a ball is given one complete rotation, perhaps in the clockwise direction. With no slip, the inner race will travel in the opposite direction at a speed determined by the ratio of the ball diameter to the diameter of the inner race. The outer race will travel in the same direction as each ball, in the ratio of the ball diameter to the diameter of the outer race. Once this diagram is drawn, the relative distance between the elements will remain constant. And if the stationary point is moved to the outer element position, when the outer element is held stationary, it will then be possible to obtain the relative speeds of each element within the bearing.

9.3 Determination of frequency components

As an example of frequency determination, consider the bearing in Figure 9.2:

O = outer race diameter
I = inner race diameter
B = ball diameter (or roller)
N = number of rollers (or balls)
f = speed of the driven member in Hz (r.p.m./60)
C = cage diameter, $(O + I)/2$

Figure 9.3 can then be drawn, initially assuming the cage to be held and the ball given $+1$ turn (analogous to unit velocity). The location of the outer and inner

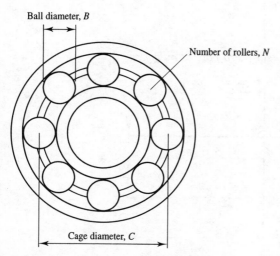

Ball diameter, B

Number of rollers, N

Cage diameter, C

Figure 9.2 *Rolling element bearing.*

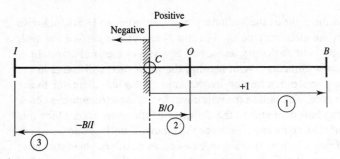

Figure 9.3   *1 ~ Initial diagram (Zero pole at cage).*

rotational speeds can be obtained by taking the relevant ratio of the ball diameter to the outer or inner diameter. The relative spacing cannot be altered.

We now shift the stationary point on the diagram to the outer location *O*, since this is now the stationary element (Figure 9.4). Proceeding with the analysis in Figure 9.5, we have made the ratio of inner to outer equal to unity and all other ratios listed; the cage diameter is used in preference to the inner and outer diameters.

The next stage (Figure 9.6) is to move the stationary point, or pole, to signify the condition where the inner race is held. The predominant frequencies that

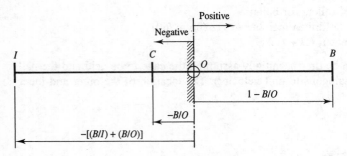

Figure 9.4   *2 ~ Outer casing held (Zero pole at outer).*

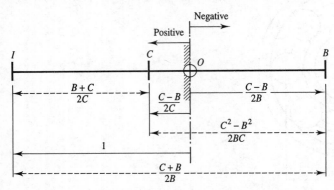

Figure 9.5   *3 ~ Outer casing held (Inner shaft speed referred to O).*

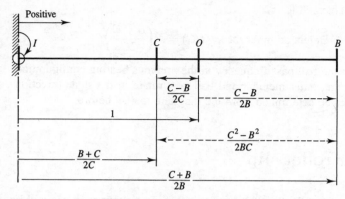

Figure 9.6 *5 ~ Inner race held (Outer race referred to I).*

appear within the bearing are taken from Figure 9.5. The formulae for these frequencies are shown below, for the condition of the outer cage held stationary.

- Defect on outer race

$$f_N(OC) = N\left(\frac{C-B}{2C}\right)f$$

(ball pass frequency)

- Defect on inner race

$$f_N(IC) = N\left(\frac{B+C}{2C}\right)f$$

- Defect on cage

$$f(OC) = \left(\frac{C-B}{2C}\right)f$$

- Defect on ball

$$f(OB) = \left(\frac{C-B}{2B}\right)f$$

$$2f(OB) = \left(\frac{C-B}{B}\right)f$$

(second-order component)

- Defect on a ball and cage

$$f(CB) = \left(\frac{C^2-B^2}{2BC}\right)f$$

$$2f(CB) = \left(\frac{C^2-B^2}{BC}\right)f$$

(second-order component)

Other gear mesh frequencies are also possible, all of which can be obtained from the diagram.

Now consider this bearing configuration: inner race held rigid and the outer race allowed to rotate with the driven shaft, as indicated in Figure 9.6. The ball pass frequency is changed to

$$\text{Defect on inner race, } f(IC) = N\left(\frac{B+C}{2C}\right)f$$

which differs from the ball pass frequency in the previous bearing configuration; furthermore, the cage, outer member and ball now rotate in the same direction. All frequencies can be determined from the diagram, just as before.

9.4   Ball or roller slip

Although the analysis of Section 9.3 assumes there is no slippage, it can still be used to give an insight into situations when slip does occur. With reference to the previous bearing, where the outer shaft was constrained, the effect of slip can be estimated by moving the stationary point away from the outer race location O. It can be established on the velocity diagram that the stationary point moves to the right, producing Figure 9.7.

Figure 9.7 shows that, for the same given input speed at I, the cage and outer race will rotate in the same direction, whereas the balls will rotate in the opposite direction. And the cage speed will increase whereas the ball speed will decrease. On a carefully monitored system, any such shifts can signify the problem of the outer cage slipping within its housing. For a bearing which has the inner race as the stationary member, both the ball and cage will rotate in the same direction with an increase in speed for both; then a frequency analyser with a zoom capacity can help to identify bearing malfunction, especially when the sideband frequencies identify the actual bearing frequency.

As an example, consider a bearing with its outer race held, outer race diameter 240 mm, inner race diameter 160 mm and 15 rolling elements of 40 mm diameter. The input shaft travels at 1200 r.p.m. and the seven frequencies are 120, 180, 8, 40, 80, 48 and 96 Hz, using the list of equations towards the end of Section 9.3. From this data the ball pass frequency is 120 Hz; but if the inner race were the stationary member, the ball pass frequency would be 180 Hz.

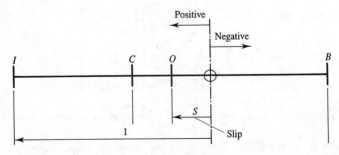

Figure 9.7   *4 ~ Outer race slipping (Zero pole shift).*

9.5 Angular contact bearings

Up to now we have assumed that the bearings have no angular contact; we have assumed that the diameter of the outer race is equal to the diameter of the inner race plus twice the diameter of the ball. But this assumption is not true for angular contact bearings (Figure 9.8). Angular contact can be incorporated into our earlier theory by using the actual cage diameter and the effective ball diameter, where

Effective ball diameter = actual ball diameter × cos ϕ

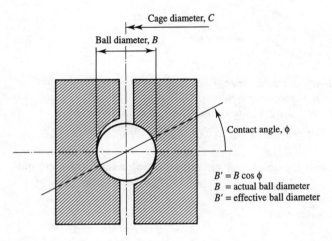

Figure 9.8 *Angular contact bearing.*

9.6 Fluid film bearings

Many machines operate on fluid film bearings instead of rolling element bearings. Fluid film bearings are used on the widest range of machines, from a dentist's drill to the largest of turbogenerators. Most machines use oil as the lubricant but an increasing number use air or the process fluid. Hydrodynamic lubrication can produce distinct problems which may often be identified by vibration monitoring.

9.6.1 Simple stability analysis

Figure 9.9 shows the action of a typical hydrodynamic journal bearing. Lubricant can be fed in at the top of the bearing but it is more common for the inlet to be situated on the horizontal axis, where the two bearing halves are usually clamped

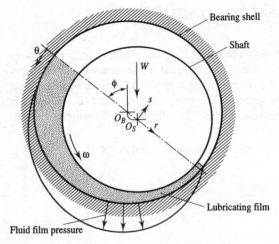

Figure 9.9   *Typical hydrodynamic journal bearing:* e = *eccentricity of shaft;* c = *radial clearance;* O_B = *centre of bearing;* O_S = *centre of shaft;* r, s *are coordinate directions;* W = *load acting on the shaft;* θ = *angle from the centreline;* φ = *attitude angle;* ω = *shaft speed;* ε = *eccentricity ratio (e/c).*

together. To analyse such a journal bearing, the following assumptions are normally made:

- The lubricant is Newtonian.

- The flow is laminar.

- There is no slip at the solid/fluid interface.

- There is no variation in pressure across the fluid film thickness.

- The film thickness is small compared to the radius of the bearing.

- The fluid density and viscosity are constant throughout the bearing.

- The journal and bearing axes are parallel.

Osborne Reynolds (1886) derived the differential equation describing the pressure generated in a thin lubricating film. This is the basis for all hydrodynamic bearing analysis. Here is a common form of the Reynolds equation applicable to journal bearing analysis:

$$\frac{1}{R^2} \frac{\partial}{\partial \theta} \left(\frac{h^3}{12\eta} \frac{\partial p}{\partial \theta} \right) + \frac{\partial}{\partial z} \left(\frac{h^3}{12\eta} \frac{\partial p}{\partial z} \right) = \frac{u}{2R} \frac{\partial h}{\partial \theta} + \frac{\partial h}{\partial t}$$

where θ and z are the circumferential and axial coordinate directions, respectively, p is the lubricant pressure, h is the film thickness, η is the lubricant viscosity, u is the sliding speed between the shaft and the bearing surface and R is the radius of the bearing. The Reynolds equation is solved for p, the lubricant pressure, in closed form for restricted situations or numerically for a bearing of finite width. There are three common approximations:

- The infinitely long bearing solution: neglecting the second term on the left-hand side.

- The infinitely short bearing solution: neglecting the first term on the left-hand side.

- The finite width bearing solution: both terms are retained using approximations to the partial derivatives (the finite difference method is the most popular).

The solution of the Reynolds equation yields the pressure around the bearing for the particular geometrical and operational conditions. Integration of the pressure field will then produce the forces generated within the bearing. Under steady operating conditions, the shaft starts at the bottom of the bearing at zero speed and, as the speed increases, the shaft centre follows a locus which approximates to a semicircle, finishing at the bearing centre for an infinite speed. The precise operating position depends on bearing geometry, lubricant viscosity, shaft speed, load, etc. A typical plot of attitude angle is shown in Figure 9.10.

Application of a disturbing force will move the shaft from this equilibrium position, altering the pressure field within the fluid and hence the forces within

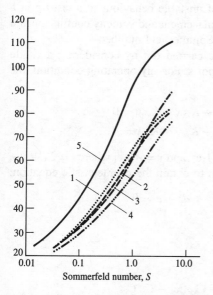

Figure 9.10   *Attitude angle versus Sommerfeld number $(S = \eta NDLR^2/Wc^2)$ for $L/D = 0.5$. (1) cylindrical, (2) 180°, (3) three-land, (4) actual bearing, (5) two-lobe.*

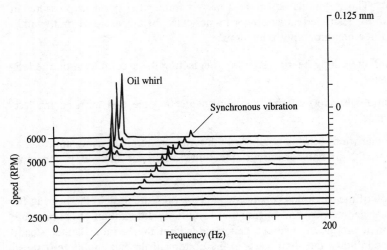

Figure 9.11 *Onset of oil whirl in a test-rig.*

the bearing. The forces that are generated normally tend to restore the shaft to its equilibrium position, but sometimes they can lead to an unstable motion or self-excited vibration. The vibration is usually at approximately 50% of the rotational frequency and is often called half-speed whirl. Half-speed whirl can be difficult to eliminate; it is often necessary to restrict the speed and/or load of the machine to avoid high levels of vibration. Figure 9.11 shows a typical waterfall plot of a rotor bearing system which exhibits the onset of unstable behaviour in a test-rig at a speed of 5000 r.p.m. Fluid film bearing displacement and velocity coefficients are shown on Figure 9.12; $L/D = 0.5$ versus Sommerfield number.

A simplified stability analysis can be carried out by considering a single bearing and the mass m of the rotor it supports. For any operating condition, the equations of motion are as follows:

$$m\ddot{x} + C_{xx}\dot{x} + C_{xy}\dot{y} + K_{xx}x + K_{xy}y = 0$$
$$m\ddot{y} + C_{yy}\dot{y} + C_{yx}\dot{x} + K_{yy}y + K_{yx}x = 0$$

Assume harmonic motion of the form $x = A_1 e^{zt}$ and $y = A_2 e^{zt}$, where $z = L_1 + jL_2$, and substitute into the equations of motion to obtain the characteristic equation:

$$az^4 + bz^3 + cz^2 + dz + e = 0$$

where

$$a = m^2$$
$$b = m(C_{xx} + C_{yy})$$
$$c = m(K_{xx} + K_{yy}) + C_{xx}C_{yy} - C_{xy}C_{yx}$$
$$d = K_{xx}C_{yy} + C_{xx}K_{yy} - K_{xy}C_{yx} - K_{yx}C_{xy}$$
$$e = K_{xx}K_{yy} - K_{xy}K_{yx}$$

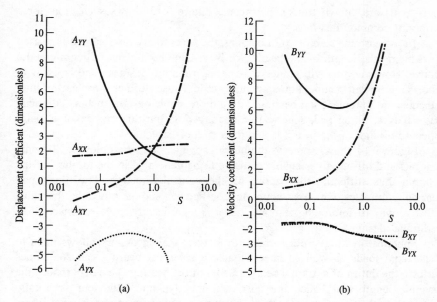

(a)   (b)

Figure 9.12   *Fluid film bearing, $L/D = 0.5$: (a) displacement coefficient and (b) velocity coefficient versus Sommerfeld number; co-ordinates X & Y.*

At the borderline of stability, z is purely imaginary i.e. $L_1 = 0$ and $z = jL_2$, so the characteristic equation becomes

$$aL_2^4 - bjL_2^3 - cL_2^2 + djL_2 + e = 0$$

To satisfy the characteristic equation, both real and imaginary terms must be zero, therefore

$$aL_2^4 - cL_2^2 + e = 0$$
$$-bL_2{}^3 + dL_2 = 0$$

From the second equation $L_2 = d/b$; this gives the frequency of vibration at the borderline of stability. Substituting this value into the first equation allows the value of mass to be calculated for operation at the stability borderline:

$$ad^2 - cbd + eb^2 = 0$$

giving

$$m = \frac{C_{xx}^2 C_{yy}d - C_{xx}C_{xy}C_{yx}d + C_{yy}^2 C_{xx}d - C_{yy}C_{xy}C_{yx}d}{d^2 - C_{xx}K_{xx}d - C_{xx}K_{yy}d - C_{yy}K_{xx}d - C_{yy}K_{yy}d + e(C_{xx}^2 + C_{yy}^2) + 2eC_{xx}C_{yy}}$$

If the mass associated with the bearing is greater than this value then instability is likely to occur; if less then the bearing should operate in a stable manner. Plots of 'critical mass' and whirl frequency can be produced if the eight bearing coefficients are known from experiments. Figure 9.13 shows such plots for a variety of common bearings.

To increase the stability of a bearing, the eccentricity ratio can be increased by reducing the length of the bearing or by machining a central circumferential groove. However, this will reduce the bearing running clearance and is likely to increase lubricant temperatures, which could cause further problems. Other remedies are to replace the bearing with a more stable design, such as elliptical, offset halves, tilting pad, spherical or floating ring, but this may involve a significant redesign of the bearings and their support structure.

Figure 9.14 shows some of the bearings which can be employed. Elliptical bearings and offset halves enable pressure to be generated in the top half of the bearing, thus stiffening the bearing in the vertical direction. The tilting pad bearing eliminates the cross-coupling coefficients and is probably the most effective way of improving stability, but it is significantly more expensive than traditional design.

Spherical bearings offer certain advantages; their geometry allows them to resist axial loads as well as lateral loads; a spherical bearing can sometimes perform the duties of a journal bearing and a thrust bearing. This may reduce the machine length and hence the cost, and the dynamic behaviour is usually improved by having a shorter rotor. There is also a reduction in frictional losses by employing one spherical bearing instead of separate journal and thrust bearings. An added benefit is an increase in stability when the bearing is carrying an

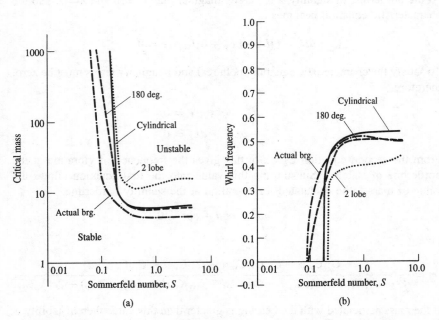

Figure 9.13   *Common journal bearings: (a) critical mass and (b) whirl frequency versus Sommerfeld number.*

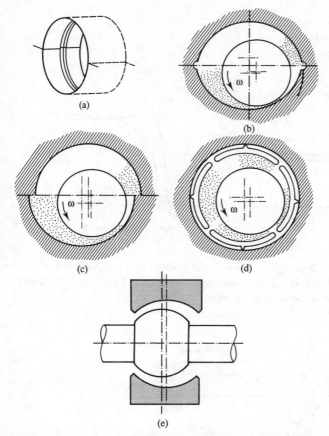

Figure 9.14 *Journal bearing designs: (a) circumferential groove, (b) elliptical or lemon bore, (c) offset halves, (d) tilting pad, (e) spherical.*

axial load. This effect is shown in Figure 9.15. Because of the freedom of motion in three dimensions, analysis of the spherical bearing requires a total of 18 displacement and velocity coefficients (Craighead *et al.* 1992).

Floating ring bearings (Figure 9.16) are sometimes used in applications where machines rotate at high speeds. The high surface velocity usually results in superlaminar or turbulent lubrication conditions, which give rise to a significant increase in frictional losses for the bearing. The ring splits the lubricating film into two, effectively halving the surface speeds for the two films. This usually results in two laminar films whose combined frictional drag is significantly less than the losses for a single turbulent bearing. The floating ring bearing usually exhibits an improved stability performance compared to a cylindrical journal bearing but is inferior to a typical elliptical bearing.

The dynamic behaviour of a bearing can change significantly if any of the basic assumptions made during its analysis prove to be invalid. This applies over

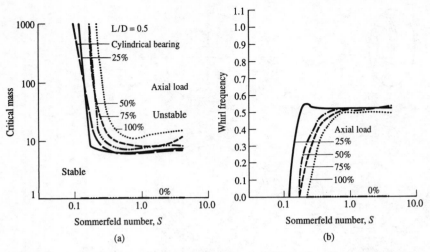

Figure 9.15 *Spherical bearing, L/D = 0.5: (a) critical mass and (b) whirl frequency versus Sommerfeld number with axial loads of 0, 25%, 50%, 75% and 100%. One trace is for a cylindrical bearing.*

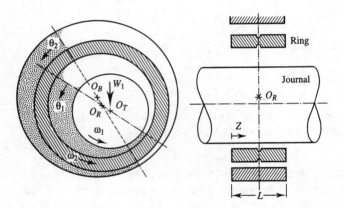

Figure 9.16 *Floating ring bearing.*

and above any assumptions made during bearing design. Two of the most important effects are non-laminar flow and journal misalignment. In general, if the lubricant flow is superlaminar, the bearing tends to be less stable than predicted by laminar analysis. Whether or not instability occurs will depend on the particular operating conditions and the degree of damping within the machine.

Both theory and experiment have shown that misalignment of the journal within the bearing sleeve can improve the stability of a rotor bearing system. It is possible to generate a pressure film over part of the top half of the bearing, depending on the orientation of the misalignment; the effect of the pressure film is similar to using elliptical bearings or offset halves. But a misalignment in bearings with washaways and jacking holes can lower stability. Washaways and jacking holes may reduce pressure generation, so the bearing operates in an unstable region.

The preceding analysis is somewhat simplified. Bearings are non-linear and instability usually produces vibrations of a finite amplitude; the size of the resulting limit cycles depends on the bearing characteristics. A machine can sometimes operate quite satisfactorily even when unstable, but most makers and users are careful to avoid it. Besides the bearings, machine stability depends on many other factors.

9.6.2 Other factors which affect stability

Seals and glands

Seals and glands have been known to produce instabilities in a wide range of machines. The need to operate with small clearances to reduce fluid leakage; this produces hydrodynamic action similar to a bearing. Figure 9.17 shows a typical labyrinth gland. Two stiffness coefficients need to be determined for seals and glands: direct (K) and cross-coupled (k); and two damping coefficients: direct (C) and cross-coupled (c). Extensive research into the dynamic behaviour of seals took place throughout the 1980s and continues into the 1990s. It has been found that a swirl brake – used to minimise the tangential flow velocity of fluid approaching the seal – is extremely effective in avoiding instabilities which arise from seals and glands.

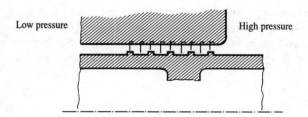

Low pressure   High pressure

Figure 9.17   *Typical labyrinth gland.*

Machine process

Instability can arise from the process within a machine. If axial flow machines are operated with the rotor eccentric to the casing, the clearance between the blade-tips and the casing will vary circumferentially. This can produce a variation in the

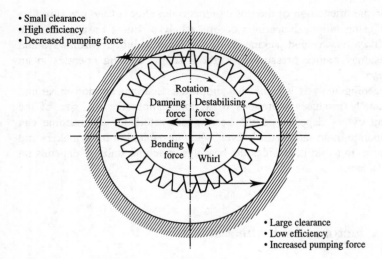

- Small clearance
- High efficiency
- Decreased pumping force

Rotation

Damping force | Destabilising force

Bending force | Whirl

- Large clearance
- Low efficiency
- Increased pumping force

Figure 9.18   *Aerodynamic instability.*

efficiency of the blading around the circumference, leading to destabilising forces, as shown in Figure 9.18.

Foundations

In many practical machines, especially large ones, the foundations provide additional flexibility, in series with the flexibility of the bearings. This will usually modify the effective stiffness and damping of the shaft supports. Theoretical studies (Leung *et al*. 1989) have shown that a more flexible foundation can generally improve system stability. However, the stiffness and damping para-meters associated with a practical foundation or support are very complex, involving direct and cross-coupled terms for stiffness and damping, as well as non-linearities, and they may be time or temperature dependent. Careful analysis should be carried out for a specific system to ensure satisfactory dynamic behaviour.

Couplings

Many large machines have several axially coupled shafts. The couplings can be designed to be completely rigid or to have a certain amount of flexibility. Investigations have shown that couplings can seriously affect rotor operation, leading to unloading of bearings and subsequent instabilities. In general, the effect of a coupling on the dynamic performance of a machine can be misleading, so a specific investigation is necessary for each particular design.

Squeeze film dampers

One means to overcome instability is to increase the amount of damping in the system. In the past 20 years, squeeze film dampers have been increasingly used for this purpose, especially in aero-engines. Figure 9.19 shows the principle of a squeeze film damper. Although their use is generally beneficial, they tend to increase any non-linear effects that are present in the system. They can also lead to jump phenomena and subharmonic resonances; this can confuse the unwary investigator. Careful design is necessary to ensure satisfactory operation of the system.

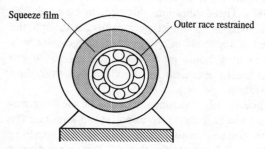

Squeeze film

Outer race restrained

Figure 9.19   *A squeeze film damper.*

Electromagnetic bearings

A recent advance is the use of electromagnetic bearings with an active feedback control loop. Aimed at eliminating rotor vibration, they apply dynamic forces which are equal and opposite to the forces generated in the machine. This technique has found success in certain specialised applications but is probably too costly and too complex for general application to rotating machines (Suzuki *et al*. 1992).

9.6.3 Conclusion

Section 9.6.2 has outlined some of the major influences on the stability of rotating machines. A rotating machine is a unique combination of many of these components and others we have not had the space to mention. Even machines which are identical at first sight may behave differently, due to variations in manufacturing and assembly tolerances, which can have a profound effect on unbalance response and especially stability. All of the effects need to be included in a reliable analysis of a rotor bearing system. The analytical methods (Institute of Mechanical Engineers 1992) are usually based on modelling the shaft flexibility using transfer matrices or finite elements, and they can include the effects of bearings, foundations, seals, glands, couplings, etc. Such analyses will usually provide information on the unbalance response and stability of the system, both of which are useful in diagnosing faults and monitoring machine behaviour. Section 11.8 is a case study which illustrates the benefits.

9.7   An introduction to gear vibration

The gear meshing and associated frequencies usually dominate any vibration spectrum measured on a typical gearbox due to the loads being transmitted through the gear teeth which are relatively high in comparison to other components such as bearings (Smith 1983). These vibrations therefore make gearboxes an obvious choice for vibration analysis to detect gear faults. Although the vibrations can usually be measured at any location within or on the gearbox, it is normal to place transducers on bearing housings to avoid resonances of the gearbox casing, and to ensure the most direct path for the vibrations from their source, the gear teeth, to the transducer.

With a well-meshed set of gears, only the fundamental gear meshing frequencies are likely to be measured and it is therefore important to know all the ratios of the geartrains within the gearbox so that each of these gear meshing frequencies can be identified (Section 9.7.1 gives the theory). But in practice, a perfect situation is unlikely to exist; instead the geartrain may be badly meshed due to misalignment or wear. Such conditions often produce secondary effects e.g. harmonics at twice or even three times the gear meshing frequencies; these sidebands are centred about the gear meshing frequencies and occur where the wear is uneven.

Secondary effects may also be caused by a breakdown of the oil film due to a loss of performance of the oil itself, or due to other local conditions such as overload. This may cause pitting on the surfaces of the gear teeth; pitting is analogous to the arcing that occurs when a welding rod is pulled away from a workpiece. These arise due to the impacting of the gears, providing a harmonic response, with vibration predominantly at twice the gear meshing frequency and sometimes at higher frequencies.

Scuffing on the surfaces of gear teeth also produces secondary effects; the scuffing alters the path of contact so the gear teeth no longer follow a true involute curve. Scuffing is usually evident when the centre distance between the mating gears is excessive, where slackness occurs within the system (e.g. an increase of bearing tolerances with time) or where the gear teeth are misaligned. If the scuffing produces an axial component of vibration as well as the more conventional radial component, this may indicate that a shaft is undergoing lateral resonance or gear misalignment.

Individual gear teeth may sometimes fracture due to fatigue or sudden overload; this will produce impacts once per gear revolution. Given the infrequent nature of the impacts in comparison with the gear meshing events, this type of fault may be difficult to observe and often requires specialist techniques such as enveloping or cepstrum analysis. Nevertheless, published results suggest that it is possible to analyse this situation using simple inputs with appropriate data acquisition and signal processing software (Maclean R. F. 1992. Gear Defect Analysis: Sonoflo Machinery Monitoring System. *Technical Report*, Sonoflo Ltd. Glasgow). An example could be a car gearbox with a tooth removed. Recent gear tooth failures have led to tragic deaths, particularly in the helicopter industry, so great importance is attached to any method which detects small defects on a single tooth.

To prolong the life of the gearbox, any significant changes in amplitude must be investigated and rectified; these amplitude changes may affect any component such as the gear meshing frequencies themselves, their sidebands or even the gearshaft frequencies. The methods used to diagnose faults may be based on frequency analysis, run-up/rundown frequency tests (Sections 10.5.3 and 11.2) or more specialised techniques such as cepstrum analysis or enveloping. Finally, remember that oil analysis can also be used for trending gearbox condition and performing fault diagnostics.

9.7.1 Elementary theory of gear vibration

The main frequencies that are likely to be generated by individual teeth or by uneven wear are the gear meshing frequency for a single contact and no loss of power. In figure 9.20:

$$f_m = w_a N_a$$
$$= w_b N_b$$

and the gear wear frequencies under the above conditions

$$f_{ga} = f_m \pm w_a \pm 2w_a \pm \ldots$$
$$f_{gb} = f_m \pm w_b \pm 2w_b \pm \ldots$$

where

$N_1, N_2 =$ number of gear teeth of gears 1 and 2

$w_a, w_b =$ speed of gears 1 and 2

And components of gear running speeds will also be present, along with their harmonics (Figure 9.21). To understand why the sideband frequencies are generated when there is uneven loading/unloading or wear on one gear tooth, assume a gear meshing frequency f_2 with a modulation (rise and fall) frequency of f_1. This frequency f_1 corresponds to the speed at which the worn gear is rotating. The overall frequency $f(t)$ can be calculated by combining the gear meshing frequency and the modulating frequency as follows:

$$f(t) = \sin f_2 \times (1 + 0.3 \sin f_1)$$
$$= \sin f_2 + 0.3 \sin f_1 \sin f_2$$

Using the standard formula $2 \sin A \sin B = \cos(A - B) - \cos(A + B)$, the equation becomes

$$f(t) = \sin f_2 + 0.5 \times 0.3 \times [\cos (f_2 - f_1) - \cos (f_2 + f_1)]$$

These relationships are illustrated in Figure 9.21.

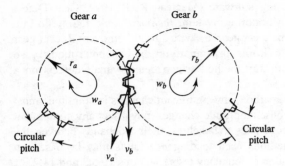

Figure 9.20 *Simple gear train.*

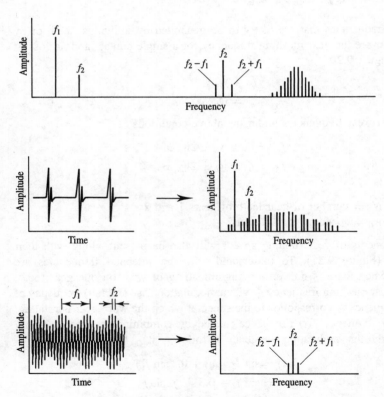

Figure 9.21 *Sideband frequencies in gear vibration:*
$$\sin f_2 \times (1 + 0.3 \sin f_1)$$
$$= \sin f_2 + 0.3 \sin f_1 \, \sin f_2$$
$$= \sin f_2 + 1/2 \times 0.3 \times [\cos(f_2 - f_1) - \cos(f_2 + f_1)]$$

9.8   Trend analysis

In a gearbox which is usually trouble-free, it may only be necessary to carry out simple severity-level tests to establish the vibration pattern. A simple integrating vibration monitoring meter is all that is necessary for this test to measure overall velocity and acceleration readings (either RMS or peak). Plotting the results over a period of time will show whether deterioration is occurring within the gearbox. Where more detail is required, maybe in critical systems, trending of relevant frequency components within a frequency spectrum can be achieved using more sophisticated commercial software.

Trending relies on a change in vibration levels over time, therefore base levels of vibration must be taken, e.g. after the gearbox has been commissioned and initially run-in. When the measured values exceed a predetermined level, usually twice the base value, the gearbox has to be investigated to establish the condition of the gears. Despite some scattering in the results due to small changes in load and speed conditions, this approach still gives a good measure of gearbox wear. Trending of simple RMS/peak acceleration/velocity readings does not give any insight into the important secondary effects (harmonics, sidebands, etc.), unlike the trending of velocity/acceleration spectra, which provides a much earlier indication of gear malfunction.

9.9   Frequency analysis

The trending of overall vibration levels, as described in Section 10.5.1, can indicate the onset of problems within a gearbox, but diagnosing the cause of the problem is not normally possible based on this information alone. This is mainly because the important secondary effects such as harmonics and bearing vibration frequencies make up only a small part of the overall vibration level; even if they change significantly, they will have little impact on the overall reading. That is why it is so important to analyse vibration measurements in the frequency domain.

Having identified an abnormal vibration trend, it is first necessary to itemise all those frequencies which are changing – the effects – and then to identify the components which are causing them to change. The following effects are typical:

- Sidebands normally appear when uneven gear wear causes the gear meshing frequency to be modulated (Section 9.7.1 describes the theory). The difference in frequency between the sideband and the gear meshing frequency indicates the rotational speed of the shaft on which the worn gear is mounted.

- Harmonics of the gear meshing frequency can be caused when the gear teeth are worn, so the loading/unloading of each tooth becomes uneven, or when misalignment of the gears causes a change in the involute path.

- Resonance of the transducer often occurs when there is metal-to-metal scuffing of the gear teeth due to a breakdown in the lubrication. Other component/casing resonances may also be excited by this noise.

- Gear rotation frequencies and their harmonics are likely to occur when a single gear tooth defects, together with component/casing resonances excited by the impacts occurring once per revolution.

Bear in mind that gearboxes generate significant levels of vibration at several frequencies (they are relatively noisy) due to the high forces present within the system. Other frequencies, due to components such as bearings, are also likely to be present (they are described in Section 9.3) and may make up a significant part of the signal, especially when the transducer is mounted on the bearing housing. This wealth, maybe surfeit, of frequency information can make it difficult to identify frequencies such as gear meshing harmonics and sidebands. The value of spectrum trending cannot be overemphasised in frequency analysis, and trending can be supplemented by using specialised diagnostic techniques such as enveloping or cepstrum analysis.

References and further reading

Craighead I.A. *et al.* (1992). An analysis of the steady-state and dynamic characteristics of a spherical journal bearing with axial loading. In *Proc. IMechE. Conf. on Rotating Machinery*, University of Bath, Sept.

Institute of Mechanical Engineers (1992). *Proc. Int. Conf. on Vibration in Rotating Machinery*, University of Bath, Sept.

Leung P.S. (1988). An investigation of the dynamic behaviour of floating ring bearing systems and their application to turbogenerators. *PhD Thesis*, University of Northumbria.

Leung P.S. *et al.* (1989). A theoretical study into the effects of damped flexible foundations on the dynamic behaviour of a 660 MW LP steam turbine. In *Proc. NATO Conf. on Vibration Wear and Damage in Rotating Machines*, University of Lisbon, April.

Lund J.W. (1987). Review of the concept of dynamic coefficients for fluid-film journal bearings. *ASME. J. Tribology*, January, **109**, 37–41.

Mischke C.R. (1990). *Bearings and Lubrication: A Mechanical Designers Workbook*. New York: McGraw-Hill.

National Aeronautics and Space Administration (1990). Rotordynamic Instability Problems in High-Performance Turbomachinery. *NASA Publication 3122*.

Reynolds O. (1886). On the theory of lubrication and its application to Mr. Beauchamp Tower's experiments, including an experimental determination of the viscosity of olive oil. *Phil. Trans. Roy. Soc.*, **177**, 157–234.

Smith J.D. (1983). *Gears and Their Vibration: A Basic Approach to Understanding Gear Noise*. New York: Marcel Dekker.

Suzuki Y. *et al*. (1992). Vibration control of a flexible rotor suspended by active electromagnetic bearings. In *Proc. Rotordynamics '92*, Venice, April.

10

Applying vibration condition monitoring

This chapter aims to provide some guidance in the setting up and running of a condition monitoring (CM) system. Although it concentrates on vibration monitoring, many of the principles can be applied quite generally. The following topics are covered:

- Deciding what equipment to monitor
- Selecting the monitoring techniques
- Setting up the measurements
- Assessing the data
- Diagnosing faults

10.1   Deciding what equipment to monitor

To maximise the benefits of condition monitoring, it is important to concentrate on problem areas; the object is therefore to assess operating equipment and grade it accordingly. Ask yourself questions like these:

- How often do failures occur?
- What are the costs of failure?
- What are the consequences of failure?

This method of setting priorities can be applied to systems (e.g. instrument air, production and fire protection) as well as to individual operation units (e.g. pumps, fans, engines). Failure and its consequences can be assessed financially by calculating average costs of failure per year, or graded more loosely into one of several categories:

- very high (severe consequences)
- medium (noticeable consequences)
- very low (minor consequences)

Grading rapidly builds a picture of the relative importance of an installation's equipment, and this can be an advantage over financial assessment. But if condition monitoring has to be justified, it is better to calculate the actual costs involved, even though it may be more difficult to obtain hard figures. Remember that the purpose is to collect sufficient data to make a decision; do not allow people to spend large amounts of effort gathering information of marginal value.

As a guideline, systems or equipment can often be analysed in the following order:

Safety systems include fire prevention or emergency power generation; they should be analysed first because they act as insurance against catastrophic failures.

Utilities may be common to all equipment; any failure would therefore prevent other equipment from functioning and might also produce secondary damage. Despite their importance, utilities are often overlooked.

Production equipment provides the tools needed to make the products or deliver the services which generate a company's profits. Without tools there is no product – and no profit. Production equipment includes many items, e.g. pumps, motors, even computers.

Ancillary equipment may be heating or lighting. Perhaps it doesn't have a direct effect on equipment functionality, but it can have a marked influence on the working environment of any operators.

Effects of failure

The effects of failure may be calculated as costs (such as spares, labour, lost production) or described in relative terms such as

- high (expensive)
- medium (moderate)
- low (cheap)

And the effects of failure can be subdivided into four main categories:

Safety/environmental risks concern damage to people or the environment; there may often be high risks in the petrochemical industry, the transport industry and the heavy manufacturing industries. Failures can lead to injury, perhaps caused by a fire or a release of chemicals.

Lost production can have serious consequences for a continuous process, such as a paper mill or a car assembly line. Stopped production may have a direct effect on output, and recovery is often difficult or expensive, perhaps requiring considerable overtime.

Secondary damage often results when a supporting piece of equipment fails, such as an oil lubrication pump. The failed lubrication pump is a small piece of equipment but its failure may damage a much larger item, causing it to need expensive repairs. Perhaps an engine will seize through loss of coolant.

Replacement/scrap costs may be significant, e.g. aviation fuel for a jumbo jet or rejects from a manufacturing plant. Replacement/scrap costs are usually higher if the equipment performs poorly; and if the equipment actually fails, spare parts and labour will also have to be costed. Specialised equipment can be very expensive to repair.

Rates of failure

In addition to the immediate effects of failure, it is also important to specify the frequency of failures – sometimes known as the failure rate. Again, this may be expressed explicitly, e.g. 10 failures per year; or it may be described more generally:

- high (occurs frequently)
- medium (occurs occasionally)
- low (occurs seldom)

Consequence of failure

The consequence of failure is calculated by combining the rate of failure with the effects of failure. For example, if a machine fails 10 times in one year, and each repair costs an average £100, then

$$\text{Cost of failures per year} = 10 \text{ failures per year} \times £100 \text{ per repair}$$
$$= £1000$$

Alternatively, we can combine the more general assessment rules into the following table:

	Effect		
Rate	High	Medium	Low
High	Very high	High	Medium
Medium	High	Medium	Low
Low	Medium	Low	Very low

It is important to remember that the presence of spares or standby units can have a marked effect on the consequence of failure. For example, a single pump may fail once every 1000 hours. Two pumps, with one in standby, would have a combined failure rate of approximately 1 in 1 000 000. Accordingly, we must reduce the consequence of failure when we have access to parallel units, standbys or spares.

Example

Here is an example of how major systems on an offshore platform were prioritised for implementing condition monitoring; note that safety was placed before production needs.

- fire water systems (pumps to provide fire protection)
- main power generation (required to run all systems)
- emergency power generation (power backup)
- instrument air systems (required for control equipment)
- gas compression modules (main 'production' equipment)
- general production equipment
- heating and ventilation

But the major systems contained many types of equipment, so each system was analysed to identify its more critical items, based on maintenance hours per unit per year and using a rough estimate of £40 per maintenance hour. For example, the instrument air systems were broken down as follows:

Type 1 generators	1400 h	£56 000
Type 2 generators	800 h	£32 000
Type 3 generators	540 h	£21 600

And the implementation costs of condition monitoring were justified by estimating the maintenance hours it would save.

10.2   Selecting the appropriate technique

There are two stages in the selection of condition monitoring techniques. First comes failure analysis to discover how a piece of equipment may fail. And then there are the methods of prevention; these too must be considered because they may also affect the choice of monitoring equipment. Ask yourself the following questions:

- What are the causes of failure?

- What are the possible effects or symptoms?

- What condition monitoring can be applied?

Where there is sufficient experience, a monitoring technique may be chosen without any formal failure analysis.

10.2.1   Analysing the failures

Failure analysis aims to identify the major aspects of equipment failure:

Modes of failure normally describe how a component fails, e.g. bearing seizure or support buckling.

Causes of failure are the reasons why a component fails; they should not be confused with the effects or symptoms of the failure. Typical causes are overload and lack of lubrication.

Effects of failure may be used to predict the onset of failure. Bearing failure, a simple example, may be forewarned by rising vibration, rising temperatures and increased noise.

Information on modes, causes and effects may come from several sources:

Breakdown history can be analysed to discover what fails, how often if fails and how long it takes to effect repairs. This information can be obtained informally using the experience of operators and maintenance engineers, extracted from supervisors' logs or retrieved from computer databases.

Maintenance manuals need to be readily accessible; they can provide an excellent source of information on failure modes and often indicate the appropriate maintenance for each one.

Table 10.1 gives some of the more obvious failure modes, plus their causes and effects; the items are listed in no special order. Table 10.2 is a typical analysis of a simple pump.

Table 10.1   *Typical modes, causes, and effects of failure.*

Modes	Causes	Effects/symptoms
Rubbing/friction	Overload	Deformation
Wear	Vibration overload	Temperature change
Fatigue	Thermal overload	Flow rate change
Decay	Misalignment	Lost efficiency
Collapse	Faulty installation	Spillages
Corrosion	Corrosive environment	Vibration
Erosion	Erosive environment	Lost performance
Debris build-up	Component age reached	Tolerance changes
Burning	Component life reached	Noise
'Failed'	Components not regularly replaced, cleaned or reworked	Pressure changes Change in levels

Table 10.2   *Typical failure analysis.*

Failure modes	Causes	Effects
Bearing collapse	Overload Imbalance No lubrication	High vibration Noise High temperature
Coupling failure	Misalignment Loose fittings Perishing	Vibration Noise Visual decay/damage/loose
Impeller erosion/fatigue	Cavitation	Vibration Noise
Filter blocked	Debris build-up	Pressure changes Visual debris
Seals worn/damaged	Perishing Wear	Visible leaks
Motor burn-out	Overload Insulation decay	High current Low insulation resistance Burning smell/temperature

This is a suggested procedure for creating a failure mode analysis; follow it for each piece of equipment.

(1) List the main mechanisms of failure.

(2) List the major causes for each failure mechanism.

(3) List the major effects/symptoms for each cause.

If a particular failure mode has more than about 10 likely causes, it is often better to subdivide it.

10.2.2 Selecting the techniques

Condition monitoring is worthwhile only if the benefits significantly outweigh the costs of its introduction. The benefits are failure prediction, failure prevention and reduced maintenance; they become obvious when assessing the consequences of failure (see Section 10.1)

Selecting an effective technique requires a good understanding of the relevant failures and the methods to deal with them. Table 10.3 indicates the relative costs of condition monitoring, and Table 10.4 matches failure types with appropriate techniques.

Table 10.3 *Relative costs of condition monitoring techniques.*[†]

Cost	Maintenance philosophy	
	Condition based	**Other**
Low	–	Wait for failure
	Manual inspections	
	Performance monitoring	Planned maintenance, easy access
	Conductivity testing	
	Vibration analysis	Planned maintenance, difficult access
	Oil/debris analysis	
	Current monitoring	Redesign, alternative supplier
	Thermal monitoring	Redesign, build in redundancy
High	Corrosion monitoring	Redesign, 'strengthen' existing design

[†] Condition monitoring costs will increase with the complexity of the technique and the frequency of the sampling. When two condition monitoring techniques have similar costs, choose the one that maximises the benefits.

Table 10.4 *Overview of common techniques and their application.*

Technique	What to monitor	What to predict	Readings
Vibration • Overall • Frequency	Gears Bearings (roller) Couplings Rotors Shafts	Imbalance Looseness Misalignment Wear Poor lubrication Cavitation Clearances	Acceleration Velocity Displacement Spike energy
Oil debris • Spectography • Ferrography • Chip detection	Bearings (plain) Bearings (roller) Gears Rotors Oil	Wear Fracture Contamination Degradation	Composition Contaminants
Inspections • Internal • External	Seals Cables Switchgear Rotating equipment	Perishing Wear Overload	Temperature Leaks Noises Burning
Current • Overall • Frequency	Motor windings Transformers	Poor insulator Worn brushes Failed rotors Air-gaps	Current
Conductivity • Overall • Transient	Motor windings Cables Switchgear	Overheating Perishing Corrosion	Resistance Capacitance
Performance • Simple • Calculated	Filters Seals Motors	Blockages Instrument drift Perishing Wear Corrosion	Pressures Flow rates Current Fuel usage

Thermal
• Spot
• Thermal
  image
	Insulation
	Bearings
	Coolant
	Lubricant
	Switchgear
	Motors
		Perishing
		Overload
		Wear
		Fracture
		Chemical
		  reaction
			Temperature

(continued)

Table 10.4   (*continued*)

Technique	What to monitor	What to predict	Readings
Corrosion • Coupons • Electrical resistance • Electrical potential	Structures Pipelines Vessels	Chemical reaction	Dimensions Voltages Resistance

This is a suggested procedure for matching the maintenance tasks to the causes of failure (Table 10.5); follow it for each cause.

(1) Identify the most cost-effective maintenance task or combination of tasks which will minimise the consequence of failure.

(2) Identify the frequency at which the tasks should be carried out based on the lead time to failures or the failure rates.

(3) Note any other causes of failure which can also use the same maintenance tasks carried out at the same interval.

Table 10.5   *Example of selecting maintenance tasks.*

Causes	Task no.	Task	Frequency
No lubrication	1	Lubricate	Weekly
	2	CM vibration check	Monthly
Shaft imbalance	2	See 2	See 2
Bearing overload	2	See 2	See 2
Shaft misalignment	2	See 2	See 2
Loosened fittings	2	See 2	See 2
	3	Inspect	Weekly
Perished couplings	4	Inspect	Monthly
Pump cavitation	2	See 2	See 2
Filter debris	3	See 3	See 3
Motor overload	3	See 3	See 3
Motor insulation decay	5	CM insulation check	Three-monthly
Failed seats	3	See 4	See 4

Table 10.6   *Example of summary maintenance plan.*

Task	Frequency	Description
1 Lubricate	Weekly	Lubricate pump and motor bearings
2 CM vibration check	Monthly	Check motor and pump for bearing overload, shaft misalignment, poor lubrication, looseness and cavitation
3 Visual inspection	Weekly	Inspect motor and pump for loose fittings, and for fluid leaks at seals Check filter pressures and motor current readings are within limits
4 Visual inspection	Monthly	Check coupling for signs of perishing and looseness
5 CM insulation check	Three-monthly	Check electrical insulation readings

This is a suggested procedure for compiling the maintenance plan (Table 10.6); follow it for each task–frequency combination in Table 10.5.

(1) List the maintenance task and frequency.

(2) List all the items to monitor and note any symptoms or effects that should be checked.

10.3   Setting up measurement points

There are three factors to consider when setting up measurement points:

- time to collect data
- time to assess condition
- effort to diagnose problems

10.3.1 Where to take vibration measurements

When deciding where and how to locate measurement points, it should be remembered that the object is to obtain

- clear signals insensitive to noise or distortion
- consistent results independent of who takes the readings
- useful data that will indicate equipment health

These principles apply to all monitoring techniques. Oil samples, for instance, should not suffer from contamination; they should be taken from the same place and should give some indication of machine condition. But for now we will concentrate on vibration readings.

Mounting considerations

Transducers must be mounted correctly to obtain clear and consistent results. Permanent mounting is the ideal method, but this may be too expensive unless the amount of access makes it hard to use a temporary mounting. Here are four principal mounting methods placed in descending order of cost:

Stud: the transducer is bolted or screwed on to the casing of the equipment being monitored; this requires one transducer per measurement point, but does give consistent results.

Snap connector: a metal fitting is screwed, bolted or glued onto the equipment at the relevant point, and readings are taken by clipping or screwing a portable transducer into the fitting. Although snap connectors produce consistent results, they may not be appropriate where access is a problem; glued fittings can break off.

Magnet: a magnet is attached to the end of a portable transducer, which is then placed by hand wherever readings need to be taken. Suitable for occasional readings or where snap connectors cannot be justified, magnets won't work on non-ferrous casings!

Hand-held probes: a pointed extension probe is screwed into the end of a transducer. Perhaps the simplest of all methods, probes should be used only when other techniques are prohibitively expensive or otherwise impractical.

The resonant frequency of the accelerometer/mounting combination will be significantly affected by the mounting method. The more secure the fixing (stud, snap), the higher the resonant frequency. Typical resonant frequencies (in kilohertz and cycles per minute) are as follows:

Stud/snap-on	> 300 000 c.p.m. (5 kHz)
Magnet	~ 120 000 c.p.m. (2 kHz)
Hand-held probe	~ 60 000 c.p.m. (1 kHz)

These resonant frequencies are often excited by wideband noise, a wide range of frequencies, often produced by rubbing, perhaps through lack of lubrication.

Cabling of transducers

The transducer wiring should minimise distortion; check for

● cable integrity (to find any breaks)

- correct earthing (to shield from interference)

- proper fixing (to avoid damage or interference)

It is important to use the correct cable when signal amplifiers are separate from the transducer. The cables are often colour coded and matched to the transducer/ amplifier combination. To avoid movement, the cables should also be secured to surfaces, perhaps with tape; movement can introduce capacitive effects which distort the signals.

The main sources of error in an integral transducer/amplifer stem from damaged cabling or connectors, quite common in portable instruments. Make regular cable inspections and always carry spares.

Location

Measurement points should be selected to obtain clear results and meaningful data. When mounting the transducers, three main factors should be considered:

Proximity: transducers need to be placed as near as possible to the source of vibration; the further away the transducer, the greater will be the attenuation. For example, if a bearing is to be monitored, ideally the transducer should be placed on the bearing housing.

Transmission: it is important that there is a clear path from the source to the transducer; the presence of a seal or flexible coupling may act as a barrier to vibrations. For imbalanced rotors, the transmission path will be through the bearings.

Orientation: the transducer should be placed to maximise the quality of its signal. Most accelerometers are sensitive in one direction only, so ensure this direction coincides with the vibration source (Figure 10.1).

There is more to selecting measurement points than where to place the transducer. The greater the number of points, the larger the volume of data to help with fault diagnosis, but the greater the time and the cost of collecting it.

- More measurement points will increase the volume of data and will make diagnosis easier.

- Fewer measurement points will cost less and will be easier to assess.

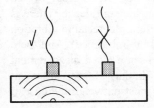

Figure 10.1   *Orientation of transducers.*

For example, the radial vibrations in the main bearing of a gas turbine are often continuously monitored using two or more permanent transducers placed at 90° to each other. A non-critical pump may have just one or two measurement points sited at each bearing.

When choosing the measurement points, remember that different components will have different failure modes and will require different monitoring techniques. Different monitoring techniques measure different variables and this may affect the placement of the measurement points.

Figure 10.2 illustrates typical measurement points on a pump; vibration analysis is used to monitor the following failures, listed in descending order of likelihood:

- bearing failures
- misalignment
- worn couplings
- loose mountings
- cavitation
- rotor imbalance

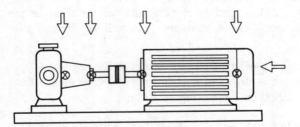

Figure 10.2   *Measuring points for pump/motor vibration.*

10.3.2 What vibration readings to take

Vibration data is used to identify machine faults which are present or are developing, faults which can be diagnosed with reasonable confidence. There is an implicit need for a link between a mode of failure and a detectable vibration characteristic (e.g. running speed frequency generated by imbalance). And there is a trade-off between the number of measurement types at a given measurement point and the cost of collecting and assessing the data. Similarly, for measurement points, the benefits of including many measurements must outweigh the cost of collecting and assessing the data.

Frequencies to cover

It is important that vibration measurements should cover the vibration frequencies generated by the components being monitored. Enough information must be available to monitor condition, and it is preferable if faults can also be diagnosed without taking extra readings in greater detail. The main danger is taking measurements which are insensitive to major modes of failure. For example, overall vibration readings may not have sufficient resolution to identify some bearing faults. Here is a general guideline to some of the more important frequencies that should be covered using vibration monitoring:

Running speeds and harmonics: running speeds are typically in the range 1500–4500 c.p.m. (25–75 Hz), and may generate up to five harmonics, typically caused by misalignment and imbalance.

Component specific frequencies: many components, such as gear teeth, fan/pump blades, ball-bearings, belt drives and couplings, will generate specific frequencies which can be monitored to assess component condition. The frequencies of interest can be calculated by multiplying machine running speeds by the number of gear teeth, the number of vanes, the number of bolts in a coupling, half the number of ball-bearings, and so on. Also of interest are sidebands of $\pm \omega$, the machine running speed; sidebands indicate modulation of the signal or beating, perhaps due to uneven gear wear.

Probe resonance: wideband noise, perhaps generated by poor lubrication, excite the resonant frequency of the probe/accelerometer; it can therefore be used as an indicator of component condition. Wideband noise can be in the range 60 000–300 000 c.p.m. (1–5 kHz) depending on the mounting method (discussed earlier). Remember that frequencies above the probe resonance will be attenuated, so try to choose a mounting method which places the resonance above component-specific frequencies.

Low frequencies: sometimes it may be useful to eliminate unwanted low frequencies (e.g. 120 c.p.m.), possibly caused by hand-shake or soft mountings. This prevents the useful component-specific data from being swamped by background signals.

On a bearing, we may be interested in the running speed to indicate imbalance, ball passing frequencies to indicate bearing faults, and probe resonant frequency to highlight lubrication problems. Measurements taken using a hand-held probe need to be filtered through a high-pass filter of 120 c.p.m. (2 Hz).

Types of vibration reading

Vibration readings fall into two broad categories:

Overall readings cover a range of frequencies (e.g. 2–10 000 Hz). Often derived using hardware, e.g. filters or integrators, they give the results as an overall signal level in real time. Overall readings are quick to take and quick to monitor, but they can be insensitive to small changes at a particular frequency.

Frequency (spectral) plots show individual frequencies which make up a signal, often plotted as amplitude against frequency. They can be derived using hardware or software, but software is normally cheaper to implement. As a general rule, the faster the results are required, the more expensive the equipment. Spectral plots provide significant resolution, but the data may take longer to collect and interpret than for overall readings. It is common to average several results to reduce the impact of noise, but this also increases the time spent on gathering data.

Types of vibration measurement

Vibrations measurements are normally readings of acceleration, velocity or displacement taken using accelerometers, velocity transducers or displacement probes. Displacement readings can also be obtained from velocity probes and accelerometers using one or two integrating circuits, respectively; velocity readings can be taken from accelerometers using a single integrating circuit. Here is a brief comparison:

Acceleration gives a relative indication of forces acting on a system (force = mass × acceleration). For a given frequency, acceleration amplitude $= 2\pi \times$ frequency × velocity amplitude.

Velocity gives a relative indication of energy dissipation (or work done) at a given frequency. Velocity amplitude = acceleration amplitude/$(2\pi \times$ frequency).

Displacement gives an indication of movement (force = displacement × stiffness); it is very sensitive to lower frequencies.

Measurement readings will be processed to give overall values, frequency spectra or a combination of the two. Remember that more information simplifies diagnosis and makes detection more reliable, but it increases the cost of collection and assessment. Here are some examples of signal and process combinations, including a specialised technique using spike energy that was developed by IRD Mechanalysis of Colombus, OH.

Overall velocity gives good sensitivity to lower frequencies up to a limit of around 3 kHz, which makes it useful when monitoring machine running speeds or harmonics, and when referring to standards (e.g. BS 4675) where magnitudes are often quoted in mm s^{-1}. Limitations of the transducer and other instruments may restrict the response to below 2 kHz, possibly less than the probe resonant frequency.

Overall acceleration is quick and simple to take. A good general measurement, it covers frequencies up to and beyond 20 kHz and is suited to monitoring gear teeth, blade passing frequencies and probe resonances. Overall acceleration tends to be insensitive to certain frequencies that are small compared to the overall signal level, and its frequency response is often limited by factors such as probe resonance, or frequency response of transducers and instrumentation.

Spike energy is an overall reading developed by IRD Mechanalysis; it is similar to Endevco's high-frequency detection system, Hf D (Endevco UK Ltd, Royston). Covering the frequency range 5–60 kHz, it can investigate higher resonances, perhaps caused by lack of lubrication, but can also respond to cavitation and gear noise. The elevated frequency band means that it may not pick up the fundamental probe/transducer resonant frequency.

Velocity spectra provide good sensitivity to lower frequencies in the same way as overall velocity readings. The preferred method is to set a frequency range, which includes the probe resonance if it lies below 2 kHz, while ensuring sufficient resolution to identify important frequencies. On a digital system, a frequency range of 2–2000 Hz with 800 spectral lines and 4 averages would give a resolution of 2.5 Hz, sufficient for machines running at 25 Hz or more.

Acceleration spectra are more appropriate for higher frequencies (> 500 Hz) generated by components such as gear teeth and gas turbines blades. Consider a gearbox with a toothed gear having 50 teeth and running at a frequency of 500 Hz; it would generate a frequency of 2.5 kHz with sidebands of 50 Hz. To obtain spectra over a higher frequency range but without a decrease in the bandwidth will require a corresponding increase in the number of spectral lines. For simpler machines, acceleration spectra may not provide significantly more information than velocity spectra, and acceleration spectra may have lower resolution.

Overall readings from spectra may be obtained on some digital systems by combining several frequency components. For example, the RMS value could be extracted for frequency values at and near to a gear meshing frequency. Once one spectrum has been obtained, several overall readings may be extracted without increasing the data collection time; this makes monitoring much simpler. The main problems include a significant set-up time and difficulty in coping with machines that have variable speed.

General vibration techniques

Several specialised techniques are excellent at helping to monitor and diagnose specific problems, but because they are specialised, they may not be appropriate for general monitoring situations. Here are a few of them:

Kurtosis is an overall reading that indicates the spikiness of a signal by giving more weight to amplitudes greater than the RMS; it measures $\text{amplitude}^4/\text{RMS}^4$. Kurtosis is typically used on bearings.

Shock pulse was developed by SPM Instruments of Bury. It evaluates shock pulses 'caused by mechanical impacts on rolling element bearings' by giving an overall level of shocks against the background signal; this identified faults such as poor lubrication, poor installation and bearing damage.

Enveloping plots the frequencies present after filtering and application of a signal follower. This makes it easier to see the defect of bearing and gears.

Cepstrum analysis provides information about frequencies which repeat themselves in a frequency spectrum. It does this by taking the log of the amplitudes – to increase the significance of lower amplitudes – then reconstructing a second frequency spectrum (see the Bruel & Kjaer application note *Cepstrum Analysis and Gearbox Fault Diagnosis*).

Table 10.7 is a brief example of the type of measurements that would be taken monthly on a non-critical, centrifugal pump running at 1500 c.p.m., using a hand-held probe.

These measurements would be taken to look for bearing damage, misalignment, worn couplings, loose mountings and cavitation. The overall acceleration and overall velocity would pick up poor lubrication, and would be confirmed by looking at the velocity spectra at probe resonance. Misalignment and loose mountings would be covered by the overall velocity and spectra, with

Table 10.7 *Measurements from a centrifugal pump.*

Measuring point	Measurement
Motor: non-drive end, vertical	Overall velocity Overall acceleration Velocity spectra 2–2000 Hz, 800 lines, 4 averages)
Motor: non-drive end, axial	Overall velocity
Motor: drive end, horizontal	Overall velocity Overall accleration Velocity spectra (2–2000 Hz, 800 lines, 4 averages)
Pump: driven end, horizontal	Overall velocity Overall acceleration Velocity spectra (2–2000 Hz, 800 lines, 4 averages)
Pump: non-driven end, vertical	Overall velocity Overall acceleration Velocity spectra (2–2000 Hz, 800 lines, 4 averages)

misalignment being identified by high readings of axial velocity. Coupling problems would be highlighted through rising overall velocity and spectral readings at a frequency specific to the coupling. Cavitation could be confused with bearing damage, but could be identified by using additional velocity readings on the pipework.

10.3.3 How often to take readings

To predict the onset of failures, condition monitoring must collect sufficient data then carry out appropriate assessment to provide enough warning so there is time to take corrective action. Obviously we do not want to expend too much effort collecting the data if no benefits will result. Consider the following factors when deciding how often to take readings:

- rate of deterioration
- cost of collecting data
- consequence of failure
- time to assess the data

The rate of deterioration depends very much on the type of component and the mode of failure:

Wear-out components tend to last a fixed length of time; they are usually items subject to mechanical wear, such as brake pads or seals. Checks are made to highlight any changes in rate of wear and to act as a guide. At least five readings should be taken before a component is replaced; a planned approach may be more appropriate than a condition-based approach.

Early-life components tend to fail very soon after their first use or else they survive. Characteristic of electronic items, such as computers, this behaviour is seen in most components to some degree. Equipment under test is run continuously for a short period when first installed; early-life components are those which fail during this test. Once this test has been completed, then measuring intervals should be chosen as appropriate for either wear-out or random dominated characteristics if it is appropriate.

Random components do not appear to fail in a predictable manner; this tends to be true of most equipment, and is often affected by environmental factors such as loads, exposure to elements, initial installation and service intervals (perhaps for lubrication). Because the failure rate is unpredictable, measurement intervals are dictated by the rate at which a failure develops:

> **Instant failure** is typically caused by overload and fatigue. There is very little warning of failure, so continuous monitoring is required.

> **Gradual degradation** is typically caused by wear, perishing and corrosion.

The warning effects develop over days, months, even years, so readings need to be taken daily, monthly, etc. Their frequency will depend on whether or not an item is critical and the degree of sensitivity with which any deterioration can be measured.

The following table is a guide on how often to take vibration readings. If the cost of collecting the data at the selected frequency exceeds the benefits, consider using a different monitoring technique or an alternative approach, perhaps providing spares. In the table, a 'high' failure consequence applies to unduplicated capital equipment, 'medium' applies to duplicated capital equipment or unduplicated low-cost equipment, and 'low' applies to minor equipment where spares are available.

Type of failure	Consequence of failure		
	High	Medium	Low
Wear-out	Monthly	Three-monthly	Twelve-monthly
Early-life	Three-monthly	Twelve-monthly	Not applicable
Random			
Instant	Continuous	Daily	Not applicable
Gradual	Weekly	Monthly	Three-monthly

There are some important points to remember when using this table. Failures can be minimised by adjusting the measurement frequency through the experience gained by operators. If failures are occurring but they are not being identified, the data collection needs to be modified: adjust the rate or choose a new technique. If equipment begins to deteriorate, take readings more frequently to provide more detailed information and to highlight an increase in the deterioration rate. The time between collecting the data and assessing the information must be at least half the time between taking readings; data is useless if it is not available to warn of failure before the next readings are taken. And remember that continuous monitoring systems generally cost significantly more than using portable data collection systems, sometimes by a factor of 10.

10.4   Assessing measurement data

10.4.1 Setting alert levels

The purpose of setting alert levels is to warn of changes in condition, so that potential faults can be highlighted and corrected before any failure occurs. Many systems rely on alerts to be generated by the monitoring system rather than requiring all parameters to be assessed continuously. But the alert levels must be set correctly. If the alerts are too insensitive, failures may occur without much warning; if the alerts are too sensitive, alerts may often be generated without real cause, wasting time and lowering confidence in the monitoring.

Alert levels are set for many types of parameters:

- Increasing, e.g. vibration signals, temperatures.

- Decreasing, e.g. supply pressures, flow rates.

- Increasing and decreasing, e.g. differential pressures, electrical current.

Alert levels need to be set for each individual parameter, taking into account factors such as equipment location, operating duty and environment. Two identical pumps in separate locations may give very different vibration readings, perhaps due to a difference in mountings or variations in pump-to-motor alignment. It is therefore important to use alert levels to highlight *changes* from the norm. Two distinct levels are normally used to indicate a change away from a norm:

Warning levels indicate a change away from the normally operating value. The best way to set them is to take at least five readings over a period of time, and to set the warning level within 10% of the average of these (as long as the readings are not rising). This may not be possible when setting the alerts for the first time, so use experience or an appropriate standard (e.g. VDI 2056, Table 10.8); reset the alerts as soon as a better norm can be determined. Remember that levels given by standards tend to be very general, they are more suitable for groups of machinery instead of individual parts of a machine, and rigidly applied, they may produce warning levels that are too sensitive or too insensitive.

Alarm levels indicate the maximum allowable value of a parameter; they are often very difficult to set for vibration readings since the maximum tolerance to vibration will depend on the strength of the equipment, and this will depend on its design. As a practical guide, vibration alert levels can be set at between 2 and 4 times the normal operating level or at the 'unsatisfactory' level given in vibration standards, using whichever is the lowest. This acknowledges the approach used in

Table 10.8 *Example of VDI 2056 vibration standard for medium-sized machines.*

Vibration level: RMS velocity (mm s^{-1})	Assessment level
Below 1.12	Good
1.12–2.8	Satisfactory
2.8–7.1	Just satisfactory
Above 7.1	Unsatisfactory

many standards, and often borne out in practice, that a doubling in vibration value indicates an increase in severity level. For comparison, parameters such as motor current alarms are relatively simple to set by obtaining the maximum current rating, often indicated on the casing plate.

Warning levels need to be reset on a regular basis, and the first reset is especially important. Judicious resetting helps false positives and false negatives.

10.4.2 Assessing changes in condition

Condition monitoring data needs to be examined for any departures from the norm – their instantaneous magnitude and their rate of change. Then any problems can then be prioritised according to their severity, and corrective action according to its urgency. The alert levels should have been set to allow changes from the norm to be highlighted automatically, either by generating reports in software or triggering hard-wired alarms. Before assuming that a fault has been highlighted, consider the following questions:

Is it a false alert? The first step should be to determine whether the alert has been triggered because the alarm level is too sensitive. This is achieved by obtaining a plot of the readings over a period of time. If no clear trend is present, and the current reading that has triggered the alarm is within the normal spread of readings, the alert level should be reset accordingly.

Is the reading correct? In cases where a single reading shows a sudden departure from the normal range of readings, the validity of the reading should be checked by retaking a measurement. These types of error may arise from faulty instrumentation, such as damaged cabling; they may also occur because manual readings are taken by different people on separate occasions.

Having observed an abnormal trend, decide on the action to be taken. The main objective is to take corrective action before any failures occur. Being too quick to react and initiating repairs too soon may be wasteful of limited resources, but being too slow to react may result in expensive breakdowns. So the main question to ask is: How soon must corrective action be taken? There are four main factors to consider when predicting when to take action.

- the current reading
- the rate of increase in the reading
- the maximum allowable reading
- the time to diagnose and correct the problem

The maximum allowable value should be taken as the alarm level (e.g. 2–4 times the normal reading). Both the instantaneous vibration level and its rate of increase will be known, so it should be possible to extrapolate the readings and thus estimate the time taken to reach the alarm level. The time taken to diagnose the problem will depend on experience and organisational limitations, and both should be taken into consideration. Remember also that some readings increase in an exponential manner – their rate of increase also increases.

Many condition monitoring systems can generate several alarms at the same time; this approach allows the response to be prioritised. In systems where readings are taken periodically, the question: How soon must corrective action be taken? becomes: Does action need to be taken before the next set of readings are collected? Readings may be collected more frequently in order to monitor the equipment more closely and to make a more effective judgement about when action is required.

Figure 10.3 is a simplified example of vibration trends over time. The warning level has been set just above the normal spread of readings, and the alarm level has been set to approximately 4 times the normal value. An upward trend is detected when the readings rise above the warning level, point A. The dashed portion of the graph from A to B extends a best-fit curve to reach the alarm level, often performed mentally. The time taken for the readings to go from A to B indicates the urgency of the problem. If this time were 1–2 days, instead of 1–2 months, the problem would need to be diagnosed and corrected quickly.

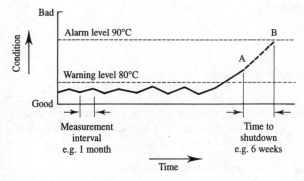

Figure 10.3   *A developing trend.*

10.5   Diagnosing faults

Fault diagnosis has the following objectives:

- to locate the component at fault
- to find it in the shortest time
- to find it in the most cost-effective manner
- to effect a solution

There are two main methodologies for linking cause and effect:

- replacing components until the problem is eliminated

- aquiring information to predict and eliminate the problem

Practical strategies combine them in various ways, according to the equipment under analysis. A gas turbine will require more data to be gathered and analysed because it is rarely feasible to swap components – they are too costly and it is difficult to disassemble a turbine. But a simple pump or fan may be most easily repaired by replacing a drive belt or coupling.

Here is a general diagnosis procedure:

(1) Identify and state clearly the effects which indicate the problem; locate the position which generates the strongest effects.

(2) List the major causes of component failures which would produce the effects at the location identified.

(3) To confirm or reject each cause, identify one or more of the following actions, listed in order of preference:

    (i)   a simple measurement

    (ii) a simple inspection

    (iii) a simple test

    (iv) a simple component replacement

(4) Carry out the confirmatory actions, starting with the simplest, until causes can be ranked in order of likelihood so that a diagnosis can be made with confidence.

(5) Carry out the corrective action and confirm the solution.

There is always a conflict between collecting enough information to allow fault diagnosis and collecting more information than can be assimilated. In general, it is better to start with a little information which can be reviewed easily; further information can be obtained as required.

10.5.1 Trend analysis

Overall trending

Trending of overall vibration readings is the first step in identifying a change in machine condition (Figure 10.4), and sometimes in pinpointing a fault location. A

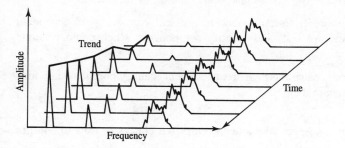

Figure 10.4 *Waterfall diagram for frequency analysis.*

rise in vibration on only one pump bearing in the radial direction would suggest a bearing problem, whereas a rise in vibration at bearings on either side of a coupling would suggest a fault in the coupling. But trending of overall vibration readings can rarely diagnose a specific problem; it cannot discriminate between individual frequencies, so it must be backed up with more information.

Spectral trending

Trending of vibration spectra examines changes in amplitudes of particular frequencies. By identifying the frequency component that is altering, it is then possible to suggest the component which is causing the problem and the particular defect. A rise in the frequency component from the outer race of a bearing would indicate a race defect or a significant change in lateral loading.

The idealised spectral trend in Figure 10.4 shows several frequencies with wideband at the top end of the frequency range and an increasing level of vibration at the running speed.

Trending of spectra can be more time-consuming than overall trending, especially if carried out manually. Some software allows trending of specific spectral bands. This allows a compromise between the intensity of spectral analysis and the lack of detail of overall signal analysis, a compromise that improves the effectiveness of fault detection and diagnosis.

10.5.2 Frequency analysis

Frequency analysis links a problem frequency pattern, the effect, to a particular component, its cause. If this can be achieved with confidence, then a diagnosis can be suggested. Here is a procedure which follows the general procedure on page 171:

(1) Identify any problem frequencies that are being generated.

    (i)   If wideband noise, investigate wear, rubbing or lubrication.

    (ii)   If one or more discrete frequencies, check if they are speed dependent (use a rundown test) and identify components which could generate them.

    (iii)   If sidebands, calculate their frequency and identify components which could generate them.

(2)   Having identified suspect components, carry out checks or tests to confirm the diagnosis.

A rise in a broad frequency range at a transducer resonance on a bearing housing suggests the occurrence of some form of rubbing, perhaps caused by severe wear or a lack of lubrication. Lack of lubrication can easily be verified by appropriate greasing or by inspecting the races during shutdown. Such actions will help to confirm or reject possible causes.

    Suppose a gearbox is generating a frequency higher than normal: 1 kHz with sidebands of 20.4 Hz. The input shaft rotates at 50 Hz, the input gear has 20 teeth, the output gear has 49 teeth, giving a gear meshing frequency of 1 kHz (20×50) and an output gear speed of 20.4 Hz ($50 \times 20/49$). This would suggest a gear wear problem on the output gear.

    Table 10.9 is a guide to identifying frequencies and problems; it should be used in conjunction with Table 10.10. Some frequencies may overlap – e.g. a gear meshing frequency may overlap a bearing frequency – so it is important to carry out cross-checks.

10.5.3 Diagnostic tests

Impact tests and run-up/rundown tests are often useful in differentiating frequencies which are dependent on running speeds from system resonances, which remain constant with equipment running speed.

Impact tests

Impact tests are often used to obtain resonant frequencies, normally of stationary structures such as a piping systems, steel frameworks or vessels. The method is to strike the structure with a hammer then to examine the response of the system using an analyser. The analyser is synchronised to collect data at the time of impact, and this is normally achieved by setting its trigger level.

    Impact analysis works because the hammer blow generates a wide range of frequencies; some of them excite the structure's natural frequencies, which are then recorded by the analyser. When a bell is struck with a hammer, it rings at its natural frequency, all others are attenuated.

Table 10.9 *Identifying frequency problems.*

Problem frequency	Direction	Faults	Possible checks
wideband noise at transducer resonance (2–10 kHz)	Radial	Lubrication Cavitation Rubbing	Try lubrication Check for noise on pump casing
1 × running speed with lesser harmonics	Radial	Unbalance Looseness Resonance	Try tightening, loosening or inspecting the holding bolts Check whether the resonance frequency is constant on rundown
50 Hz	Any	Electrical	Disappears if machine is turned off
running speed/ 2 × no. of rolling bearing elements	Radial	Bearing failure	Lubricate and see whether problem disappears; visually inspect if possible
running speed × no. of teeth + sidebands	Radial	Gear meshing Looseness	Visual inspection of gears
2 × running speed with lesser harmonics of 1×, 3×	Radial + axial	Misalignment	Check whether vibration occurs in the axial as well as the radial direction
running speed × no. of fan blades/pump vanes	Radial + axial	Blockages Incorrect tolerances	Check flow rates and temperature rises
running speed × no. of coupling bolts	Radial + axial	Loose coupling Misalignment	Inspect coupling bolts

Table 10.10 *Machine faults listing.*

Probable causes of vibration	Dominant vibration effects		
	Frequencies	Direction	Comments
Unbalance	Incorrect mass distribution	1 × r.p.m.	Radial	May develop due to uneven wear or build-up of debris. Shaft whirl can sometimes be mistaken as unbalance.
Misalignment	Component centres offset and/or at an angle, or distortion of casings	2 × r.p.m., 1 × r.p.m. with multiples	Axial and radial	Sometimes confused with looseness; the main difference is the presence of an axial component.
Looseness	Large clearances in bearings, gearboxes; loose mountings	1 × r.p.m., 2 × r.p.m. with multiples	Radial	Flexible mountings can give similar effects which may be considered normal. Use trending to monitor changes.
Bearings	Lack of lubrication Rubbing due to failed rolling elements or incorrect tolerances	broadband, inducing transducer resonance	Radial	Broadband noise can also be caused by cavitation. Take readings on pipework to help differentiate them. The vibration level may drop if bearings fail and the shaft/housing behaves as a plane bearing.

Bearings (*continued*)	Whirl on plain bearings	$0.4 \times$ r.p.m.–$0.5 \times$ r.p.m.	Radial
	Defects in rolling element bearings	no. of elements $\times$ r.p.m./2 for race defects	Radial	Cage and element defects can also occur. Some larger bearings can be visually checked for surface condition.
Couplings	In-line: perished rubber or loose coupling bolts can give problems	r.p.m. $\times$ no. of bolts/ contact points	Radial	It is often simple to check the condition by visual inspection.
	Belt	$2 \times$ belt r.p.m. or $3 \times$ belt r.p.m.	Radial	Can be checked visually using strobe or checking slack. Toothed belts may generate a high frequency (no. of teeth $\times$ r.p.m.).
Gears	Uneven tooth wear	r.p.m. $\times$ no of teeth with sidebands 1 or more $\times$ r.p.m. of worn gear	Radial	
	Increased tolerances on teeth or supporting bearings	r.p.m. $\times$ no. of gear teeth	Radial	Gear meshing frequency is to be expected. Problems will be seen as a rise over time Epicyclic boxes are sensitive to changing mechanical tolerances.

(*continued*)

Table 10.10 (*continued*)

Probable causes of vibration	Dominant vibration effects			
	Frequencies	Direction	Comments	
Resonance	Casings, structural supports, components	$1 \times$ r.p.m. or multiple of some excitation frequency	Radial and axial	Resonance is caused by some other frequency, and disappears quickly if the running frequency changes (e.g. when turning off the machine).
Aerodynamic and hydrodynamic forces	Incorrect tolerances between casing and rotors	r.p.m. $\times$ no. of fan, pump or compressor blades	Radial	Blade passing frequencies are often present, but should not be excessive and should not significantly change if a unit is rebuilt
	Cavitation	broadband	General	
Electrical	Thyristor speed control	multiple of line frequency	Radial	Will disappear immediately the power is turned off.
	Stator problems, loose bars	$2 \times$ line frequency	Radial	

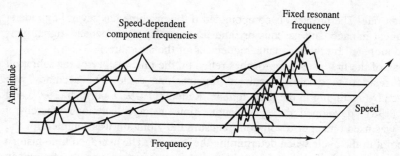

Figure 10.5 *Results from a typical run-up test.*

The size of the hammer depends mainly on the size of the structure under test. A heavy paper mill would require a sledgehammer, whereas a tiny bell would need only a small hammer.

Run-up/rundown tests

Run-up/rundown tests are normally performed on dynamic systems in which the speed of the machine can be varied; they are used to distinguish between frequencies that depend on speed (such as gear meshing) and those independent of speed (natural resonance frequencies). The method is to take a number of spectra in sequence as the equipment speed is increased. This produces a series of spectra from which to determine the fixed and variable frequency components (Figure 10.5).

Run-up/rundown tests allow the frequencies generated by the equipment to be produced under normal operating conditions, and the resonance frequencies can be quickly differentiated from speed-dependent frequencies. They can be used during system development to ensure that the system will not be operated at or near a resonant frequency, which could cause catastrophic failure.

10.6 How to monitor reciprocating machinery

Reciprocating machines, such as internal combustion engines, compressors and pumps, are widely used throughout industry. Fault diagnosis originally relied on analysis of the lubricating oil, on temperature monitoring and on indicator diagrams (pressure–volume traces). In the mid 1970s attempts were made to apply vibration monitoring techniques to reciprocating machinery. But despite their success with rotating machinery, the results were disappointing. The difficulty arose because the working cycle consists of many events of short duration, such as valves opening and closing or fuel ignition, which share the same basic repetition rate. The problem is compounded if the machine has several cylinders. Mechanical impacts and gaseous or fluid flows produce a wideband signal generally dominated by the resonant frequencies of the structure.

One of the first useful approaches relied on the time history of the vibration and its relation to discrete events in the machine cycle. The accuracy was improved by averaging in the time domain. Randall (1979), Courrech (1979) and Boyes (1981) developed the synchronous time average technique whereby a trigger was used once per revolution to ensure the vibration was sampled for the same point in the cycle when determining the average. The averaged time history from various parts of the cycle could then be transformed into the frequency domain, enabling changes to be observed and related to particular defects. Johnston and Stronach (1986) used a similar approach in conjunction with pressure traces and $P-V$ diagrams to diagnose valve faults in a reciprocating compressor. One of its limitations is the need to ensure a constant machine speed while the averaging is under way, or to be able to compensate for speed fluctuations. Enveloping the vibration signal in the time and frequency domains can provide clarification of data for diagnosis and is useful in discriminating between speed-related problems and structure-related problems.

A development of synchronous averaging, the gated analysis technique, also samples the signal using a trigger once per revolution but it has a variable time delay; this means that spectral analysis can be carried out on the signal a small section at a time. Repeating the analysis with increasing delays allows the whole cycle to be investigated piece by piece. Data is usually presented in a waterfall plot, each spectrum based on a small crank-angle segment of 5–45°. Examples of the technique applied to diesel engines include Courrech (1989), Serridge (1991) and Leong and Lim (1992). The disadvantage of gated analysis is the amount of data collected and the time it takes to be analysed. But the use of PC monitoring systems and the development of dedicated software can automate the process to some extent, allowing gated analysis to form part of an integrated monitoring programme. This extension of gated analysis technique is called cyclic analysis; it is described in more detail by Serridge (1991).

10.7 Practical condition monitoring procedures

We now consider selection and acceptance procedures for the purchase of rotational machinery, but only those procedures or items of equipment that are currently available to industry.

10.7.1 Commissioning tests

A machine frequency spectrum in the form of a waterfall diagram should be produced on commissioning; this ensures the system has no major resonant frequencies within its operational speed range. A useful analogy is the conventional washing-machine with spin-dryer. During run-up the machine behaves well and during initial rundown it behaves normally. However, the machine shakes violently at a certain speed during rundown, and prolonged operation at this speed would quickly reduce the machine's efficiency. Only a sharp drop in speed, as the vibration energy is dissipated, returns the washing-machine to normality.

This phenomenon also exists in expensive process machinery, but not usually with such spectacular results. A component that suffers from vibration may be an integral part of the machine and may not provide a significant contribution to the overall vibration waveform, but it may still fail prematurely. Although machines are seldom operated exactly at a resonance, breakdown can still occur at frequencies close to a resonance, frequencies within the operating range. The waterfall diagram for both run-up and rundown will clearly itemise such defects and verify the calculation on the Campbell diagram.

10.7.2 Periodic testing

A system should be periodically, if not continuously, monitored for any significant change in its behaviour. The rate of change of a vibratory component is more valuable than the initial absolute value of vibration. Such checks can give early warning of impending failure and give the engineer time to take avoiding action, or at least time to plan ahead so the fault can be remedied at a convenient time; this is much better than having to cope with a catastrophic failure followed by expensive shutdown. For non-essential and inexpensive machinery, a check on severity level may be all that's needed until a fault becomes apparent. Then the frequency domain equipment can be utilised to diagnose the cause of the problem.

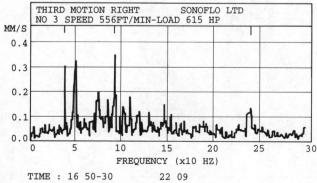

Figure 10.6 *Output from a computerised system.*

Figure 10.6 shows the output from a computer-controlled, 12-point condition monitoring system where all points are monitored in the frequency domain. Initial specimen spectrum values are stored for subsequent comparison, and an audible warning is sounded when malfunction is suspected to within prescribed limits; this gives the operator time to remedy the fault before eventual malfunction of the system by a bearing failure. The three gear meshing frequencies on the machine are flagged, leaving the other signals to be identified as bearing frequencies. This system is capable of detecting false alarms and has been working in the field for several years; the trace in Figure 10.6 indicates a bearing problem at around 52 Hz.

10.7.3 Malfunction testing

It is often possible to observe machine malfunction during normal operation. The malfunction is usually attributed to the dynamic behaviour of the machine, and although it may not affect the process greatly, it may give an early warning of a more serious machinery fault. An example is a large process-line machine which sometimes gave an audible knock at a particular speed; in fact, it was so loud it could be heard clearly throughout the plant. Many hypotheses were proposed but the reason for the knock was found from a waterfall diagram; what's more the knock could be detected even before it became audible. The results in Figure 10.7

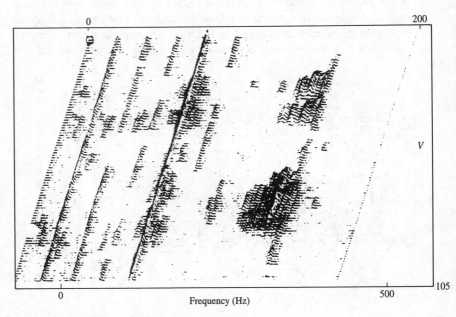

Figure 10.7   *Knock vibration test.*

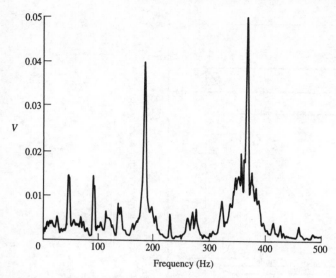

Figure 10.8 *Significant knock at 370 Hz.*

correspond to roughly $1/\omega$, where ω is the running speed of the machine, and the diagnosis shows that the system was undergoing a lateral resonance with the fan impeller striking the casing at a single place. Examination revealed that the fan impeller was striking a protuberance in the casing; the impeller had become bowed, exhibiting two slightly different bows on diametrically opposite sides. The significant knock from the major and minor bows can be seen in Figure 10.8 at around 370 Hz. This higher frequency disappears along with the knock when the system speed is altered from the resonant condition by a small amount.

10.7.4 Machine acceptance

Before a purchaser accepts any machine from a supplier, he or she should obtain verification that no major resonance lies within its range of operating speeds. A Campbell diagram (Figure 10.9) should be produced from elementary dynamic calculations. For any operational speed of the machine, this diagram clearly shows where the resonance frequencies are located (plotted horizontally at values pertaining to the resonant frequencies). It also shows forcing components at $1/\omega$, where ω is the running speed of the machine, and at any other forcing frequency. Figure 10.9 plots a four-bladed fan attached to the shaft. This will clearly show any forbidden speed range, allowing for a 10% error in natural frequency determination.

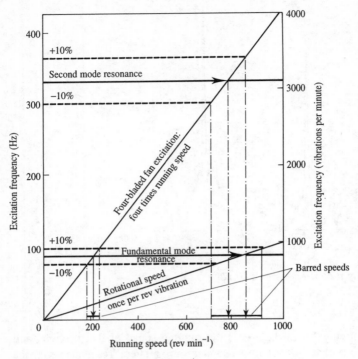

Figure 10.9   *Campbell diagram.*

References

Boyes J.D. (1981). Reciprocating machine analysis with FFT analyzers. *Bruel & Kjaer Application Note*.

Courrech J. (1979). Condition monitoring of slow-speed diesel engines. In *Proc. Motor Symposium '79*, **11**, 479–88.

Courrech J. (1989). Examples of the application of gated vibration analysis for the detection of faults in reciprocating machines. In *Proc. Noise and Vibration Conf.*, Nanyang Technical Institute, Singapore.

Johnston A.B. and Stronach A.F. (1986). Valve fault detection in reciprocating compressors. In *Proc. Int. Conf. on Condition Monitoring*, BHRA, Brighton, May.

Leong M.S. and Lim K.B. (1992). Experimental investigations on the use of gated vibration analysis for condition monitoring in reciprocating machines. In *Proc. IMechE. Conf. on Vibrations in Rotating Machinery*, Bath, Sept.

Randall R.B. (1979). Diagnostics of slow-speed reciprocating machines. In *Proc. Technical Diagnostics '79*, Karlovy Vary.

Serridge M. (1991). Cyclic analysis: an online and off-line automated technique for the vibration monitoring of reciprocating machines. In *Proc. COMADEM '91*, Southampton, July.

Case studies

The nine case studies fall into four categories

Sections 11.1 and 11.2: elementary studies.

Sections 11.3 to 11.5 and 11.8: specific applications of condition monitoring to plants and processes.

Sections 11.6 and 11.7: balancing studies.

and

Section 11.9: a condition monitoring game that simulates the monitoring of a piece of machinery.

11.1 Information from periodic waveforms

Conditioning monitoring involves the interpretation of waveforms from measurements by transducers. Most waveforms from rotating machinery will be periodic or approximately periodic. This case study is written to help with the first steps in waveform analysis using an elementary approach – direct measurement from a recording of the waveform.

However, more complicated waveforms will eventually have to be interpreted, such as recordings of transient or random signals; they require a frequency spectrum analyser (see Chapters 6 and 7).

Condition monitoring of machinery normally uses a measuring device that produces a signal which can be processed to represent a physical variable. Many of these signals are periodic or approximately periodic, so a considerable amount of information can be gleaned, even with elementary display and analysis equipment.

This is emphasised in the case study, which also indicates some of the pitfalls when dealing with periodic signals obtained from machinery. And the case study underlines the considerable range of equipment that is now available for frequency analysis.

Description of periodic waveforms for the study

Two types of periodic signals have been chosen:

- A single-frequency periodic signal of constant amplitude.

- A signal with three frequency components with different amplitudes.

Both signals are constructed from standard mathematical equations and are presented in detail. As a guide to the information on the frequency content of each periodic signal, a typical graph is shown in the form produced by a spectrum analyser.

Single-frequency periodic signal

Figure 11.1 shows a sine wave of unit amplitude with a single frequency component of 10 rad s^{-1}. The amplitude (X_{PEAK}) of the periodic wave is constant, so

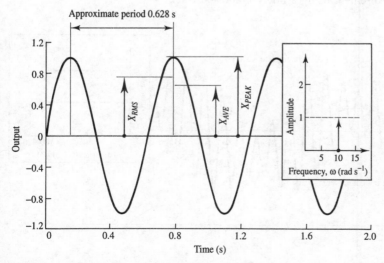

Figure 11.1   *Sine wave of unit amplitude: 1 sin 10t*

it is possible to calculate the root mean square value (X_{RMS}) and the average value (X_{AVE}) of the waveform. The frequency of vibration can be calculated by measuring the period of the wave.

Multifrequency periodic signal

A signal containing three frequency components has been constructed from the equation

$$y(t) = 5 \sin 10t + 4 \cos 100t + 8 \cos 120t$$

It is the sum of a sine wave of fundamental frequency 10 rad s^{-1}, a cosine wave at 10 times the fundamental frequency and another cosine wave at 12 times the fundamental frequency. The amplitudes of each component are different, and apart from its periodicity, the resultant waveform bears little relation to an actual machine response.

The complete waveform is illustrated in Figure 11.2; along with its first-order component. To highlight some of the difficulties with elementary analysis, the sum of first-order and third-order components is shown in Figure 11.3 and the sum of the second-order and third-order components in Figure 11.4.

Information from the waveforms

To begin with, information can be found from the single-frequency waveform by direct measurement using a finely graduated ruler.

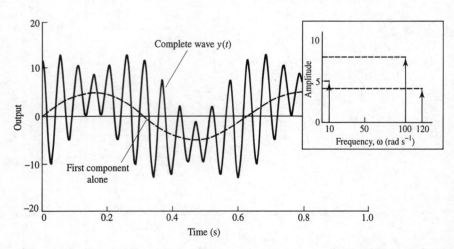

Figure 11.2 *Periodic waveform: $y(t) = 5 \sin 10t + 8 \cos 120t + 4 \cos 100t$.*

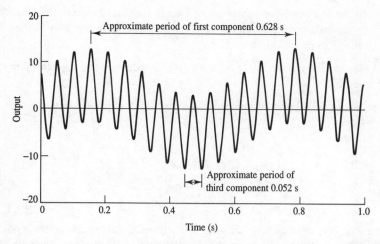

Figure 11.3 *First component plus third component of periodic waveform:* $y_1(t) = 5 \sin 10t + 8 \cos 120t$.

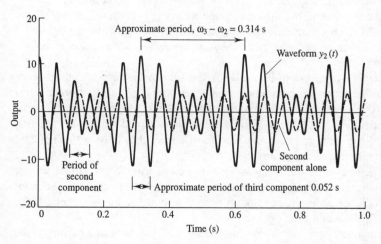

Figure 11.4 *Second component plus third component of periodic waveform:* $y_2(t) = 8 \cos 120t + 4 \cos 100t$.

The time base of the recording will normally be known accurately, so the measurement of the distances between the peaks will give the period of vibration. In this case the value is approximately 0.628 s and its reciprocal gives the frequency as 10 rad s^{-1}.

However, if a frequency analyser were available, it would display the inset graph (Figure 11.2) showing a single line at 10 rad s^{-1}. This assumes perfect analysis conditions with no effects from the electronics.

The amplitude of a waveform is always of interest; the amplitude of the wave in Figure 11.2, obtained by measurement or by peak value, is unity. Two further parameters are used to describe a waveform:

The RMS value is defined as $X_{PEAK}/\sqrt{2}$; the RMS value of the waveform in Figure 11.2 is 0.707.

The average value is related to the RMS value and the peak value by the following relations:

$$X_{AVE} = (2\sqrt{2}/\pi)X_{RMS}$$
$$X_{AVE} = (2/\pi)X_{PEAK}$$

These parameters were defined originally for use with electrical measurements of current and voltage and their use is now well established in waveform analysis.

Several questions arise when dealing with complicated periodic waveforms: (a) what are the frequency components and (b) what are their magnitudes? Figure 11.2 shows the multifrequency waveform chosen over a time period of 1 s along with the fundamental frequency $5 \sin 10t$. Some crude measurements can be made from the waveform, but the inset graph from the spectrum analyser shows how the results from crude measurements of amplitude and frequency may be misleading. This may become clearer by investigating separate parts of the complete waveform.

Consider the waveform in Figure 11.3, the sum of the first- and third-order components. Elementary measurements give the period of the first-order component as 0.628 s, equivalent to a frequency of 1.59 Hz. The period of the third-order component is 0.052 s or 19.1 Hz, close to the expected ratio of 12. A quick assessment of the amplitudes of the waveform shows that they are difficult to interpret, even in this simpler form.

The difficulties with the analysis of complicated waveforms can be further emphasised by considering the waveform in Figure 11.4. The sum of the second- and third-order components are plotted over 1 s along with the second-order component for the same time period.

Our elementary approach allows us to find the amplitude and period of the second-order component and the period of the third-order component. But we encounter one of the major difficulties when interpreting graphs of waveforms: there is a new frequency with a period equal to the difference between the frequency of the second-order component, and the frequency of the third-order component. We can also see this new frequency on the graph of the complete waveform (Figure 11.2), a frequency of 20 rad s^{-1} or 3.18 Hz.

From these examples, particularly those for the multifrequency waveform, we can see that great care must be taken to identify the constituent parts of the wave and to make sure that each frequency is due to some physical effect. The importance of Fourier analysis also becomes clear; once it could be achieved on condition monitoring instruments, all sorts of refinements were quick to follow.

11.2   Waterfall diagrams in condition monitoring

As recommended in Chapter 10, a very useful test on a machine involves controlled run-up and rundown, often monitored on a *waterfall diagram*. This process was studied on a model rotating machine.

The model machine (a Bently Nevada from Bently, Minden NV) comprises a flexible steel shaft carrying two rotors of unequal weight and supported in plain bearings. A motor is attached to one end of the flexible shaft by a synthetic rubber coupling, and the motor speed is precisely controlled by a feedback control system. Two non-contact transducers are used to monitor the shaft orbits, at an appropriate position, and either of the response signals may be monitored by a spectrum analyser.

The machine is usually run-up and rundown by a controlled ramp input to assess the most suitable speed range, then a waterfall diagram is recorded for the complete process. Before showing typical waterfall diagrams for the machine, let us consider their general format.

Basic form of a waterfall diagram

Many forms of spectral analysers have been produced (some of them by Schlumberger Technologies), and each manufacturer incorporates special facilities. However, a waterfall diagram has a basic form. Figure 11.5 shows a waterfall diagram comprising frequency spectra recorded in the form of a map plotted using three axes. The vertical axis plots the magnitude of each spectrum at a particular frequency; one of the horizontal axes plots the frequency range of the data from zero to the maximum frequency chosen in a test; the other horizontal axis, mutually perpendicular, plots the number of spectra from 1 to N for a particular map.

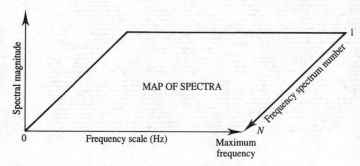

Figure 11.5   *Format of waterfall diagram.*

Typical waterfall diagram for an elementary rotating machine

Although many different machinery faults can be investigated with the test apparatus, this case study concentrates on the effects of transverse shaft vibration. The prime consideration is to find the critical speed of whirling.

The system is balanced to within acceptable practical limits, so there is no significant excitation from out-of-balance forces. Indeed the excitation is produced by the change in the system's centre of gravity caused by the flexible nature of the shaft.

From separate run-up and rundown tests, two waterfall diagrams have been recorded and displayed using the spectrum analyser. Figure 11.6 is for the following settings:

Spectral magnitude	volts peak/Hz
Number of spectra	1–45
Frequency range	0–250 Hz
Skew of diagram	none

Figure 11.6 shows the variation of the fundamental frequency during the run-up phase from a speed of 412 r.p.m. to a maximum speed of 6522 r.p.m. then on the rundown phase to the minimum speed. During run-up the first critical speed (or resonance) occurs at 54.64 Hz, equivalent to a speed of 3278 r.p.m., and on rundown this effect is clearly indicated at 59.02 Hz (3541 r.p.m.). The difference in this value of approximately 8% is indicative of the influence of weak non-linear parameters in the system, possibly the effects of the support conditions at the plain bearings.

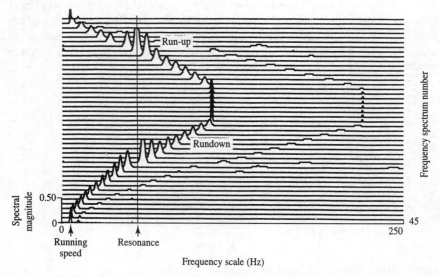

Figure 11.6   *Typical waterfall diagram for run-up and rundown.*

Figure 11.7 demonstrates some of the features obtainable on a waterfall diagram; the analyser settings are as follows:

Spectral magnitude	dB peak/Hz
Number of spectra	1–36
Frequency range	0–250 Hz
Skew of diagram	yes
Blanking of spectral lines	yes

These are very different from the settings that produced Figure 11.6; the new settings allow the diagram to be skewed and unnecessary lines may be blanked out. The run-up and rundown procedures are the same as for Figure 11.6, but the maximum speed is only 5804 r.p.m. Figure 11.7 shows that the main effect is a clear indication of the variation of the fundamental speed and its frequency components throughout the test, especially noticeable for the second-order component, which clearly follows the variations of the fundamental by a factor of 2.

The frequency variations can be clearly identified in Figures 11.6 and 11.7; this suggests that waterfall diagrams are an ideal tool for monitoring the condition of rotating machinery.

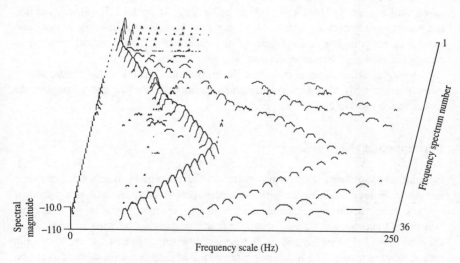

Figure 11.7 *Waterall diagram with skew and spectral blanking.*

11.3   Diesel engine and generator set problem

This case study aims to show how general vibration measurements may be applied to an engine/generator system, and to describe the link between vibration monitoring and the design of a system to withstand dynamic effects.

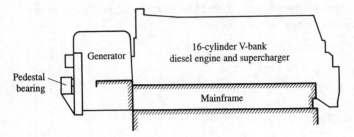

Figure 11.8   *V-bank diesel engine and generator.*

Figure 11.8 shows an outline view of a V-bank engine coupled to an electrical generator. Both machines are connected to the solid frame via suitably spaced vibration isolators. The machines have the following specifications.

The prime mover is a 16-cylinder V-bank diesel engine with a mechanically driven supercharger. The engine manufacturer states that the engine output is 1840 and 1180 b.h.p. under supercharged and unsupercharged conditions, respectively, in the speed range 750–960 r.p.m. Significantly the main torsional excitation from the engine crankshaft was reported as being at the 2.5th order at a speed of 1592 r.p.m.

The electrical generator is directly coupled to the engine crankshaft and is capable of producing 1000 kW of electrical energy. At the far end of the generator the shaft is carried in a pedestal bearing where the bearing housing is not connected directly to the main frame of the machinery. The pedestal bearing is connected to the generator endcover by a bracket.

Space considerations dictate that the engine has several bolted joints, which allow the assembly of the main parts on site; connection to the generator is made in the space available.

Description of the problem

Regulations stipulated that during operation the machinery would be monitored for vibration levels to assess whether maintenance was required. Over a period of several years, tests were carried out on many separate facilities of this type. Vibration measurements along three mutually perpendicular axes of velocity at the free end of the engine and on the pedestal bearing appeared to be above the recommended levels. The machines often required maintenance, which interfered with their operation, increased costs and reduced efficiency.

The Dynamics and Control Division of the University of Strathclyde was asked to study the system and to recommend improvements.

Vibration tests and measurements

Several tests were carried out using velocity probes (Carl Schenck UK Ltd) attached at key positions on the engine, generator and pedestal bearing. Signals from these probes were captured by a digital FM tape recorder for later analysis by software (Sonoflow Ltd) or analysed on site by a digital spectrum analyser (Hewlett Packard Ltd). The vibration tests took the following form:

- Transient vibration tests on the generator endcover and the pedestal bearing support, with the engine stationary.

- Vibration monitoring at various speeds with a 500 kW load.

- Vibration measurements at a constant speed with a 500 kW load.

- Vibration measurements during run-up and rundown tests on the machinery over the normal speed range of 750–960 r.p.m. and with a 500 kW load.

Results and recommendations

The most significant and important results were found from the transient tests and the run-up and rundown tests of the engine/generator. The transient tests identified a major resonance of the complete engine/generator on the support at approximately 32 Hz. This is close to the frequency of torsional excitation from the engine, 1592 r.p.m. or 27 Hz.

The vibration at the pedestal bearing in the vertical direction was recorded during a rundown test from the maximum speed of 960 r.p.m.; typical results are shown in Figure 11.9 in the form of a waterfall diagram. The waterfall diagram shows very significant vibration levels at the following orders of engine speed 0.5, 1, 2, 2.5 and 3.5. In particular, the 2.5th order effects are large at the maximum speed of 960 r.p.m. The effects are further clarified by displaying a single spectrum at 960 r.p.m. (Figure 11.10) where the first spectrum at 960 r.p.m. indicates that large velocity levels occur at the 2nd and at the 2.5th order components of magnitude 25 mm s^{-1} and 27.2 mm s^{-1}, respectively. The results also show significant velocity values at orders 0.5, 1 and 3.5, which also provide excitation in the engine speed range. These results indicate that the system requires changing to remedy original design errors.

During normal operation (after a suitable run-up procedure) the speed of the engine/generator is close to the maximum value of 960 r.p.m. It is clear that the problem identified by condition monitoring is due to the resonance condition at 32 Hz. When operating in the region of resonance, the whole assembly has a rocking motion on the support structure. This motion encourages the loosening of various important bolted joints, and any clearances produced are reflected in the large magnitudes of vibration at the engine and generator. This feature is viewed as a serious design fault in the machinery.

A second design fault contributes to this effect: the pedestal bearing bracket is attached to the flexible endcover of the generator. At or near resonance, large amplitudes of vibration occur at the pedestal bearing; they promote excessive wear and reduce its operating life.

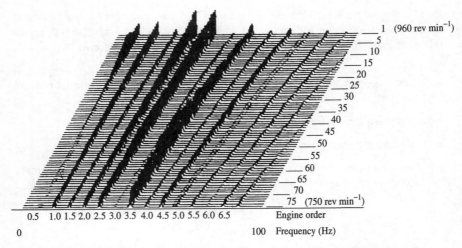

Figure 11.9 *Waterfall diagram for rundown test on engine and generator.*

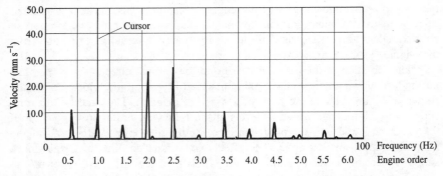

Figure 11.10 *Frequency spectrum for 960 rev min^{-1} taken from Figure 11.9.*

The following recommendations were made to improve the dynamic characteristics of the machinery and to achieve satisfactory behaviour during operation:

- A design was produced for an inflexible pedestal bearing support system, attached to the main support. This would raise the natural frequency of the machinery on the support to a value well above the resonant frequency of 32 Hz.

- Changes were recommended in the condition monitoring procedures which would take account of the 2.5th order effects.

- A range of speeds were recommended for safe and efficient operation of the engine/generator set, so that torsional oscillation effects could be minimised.

11.4   Condition monitoring at a continuous coal-handling plant

The movement of coal in large quantities requires specialised and expensive equipment. Items such as conveyors, gearboxes and loaders operate in arduous conditions; any unplanned stoppage can generate severe cost penalties, e.g. a ship may be unable to load and sail as planned.

Vibration monitoring and oil sampling/debris analysis have been applied to a coal terminal. Debris analysis has proved particularly useful for monitoring the gearbox condition and the site now has all of its gearboxes regularly sampled. Here is an example of early diagnosis.

Problem

The slew gearboxes of the coal reclaimer were sampled over a period of several years. Samples were taken by site personnel about every 6 weeks and sent to Century Oils for analysis. Some of Century's reports, called CENT reports, are shown in Figures 11.11 to 11.13. Clear indications of problems were observed, and stripdown during a planned outage revealed severe wear between the pinion

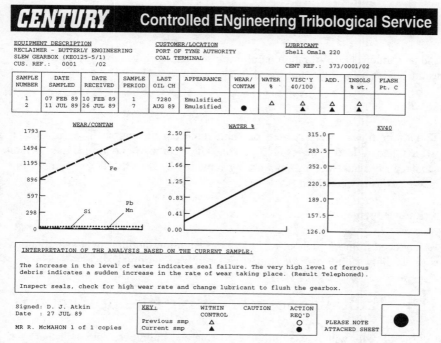

Figure 11.11   *CENT report: MORE*

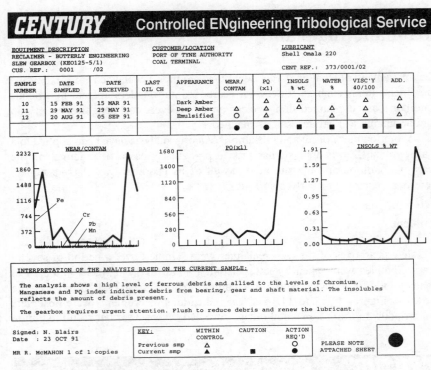

Figure 11.12 *CENT report: MORE*

and its splined shaft. Vibration monitoring showed no indication of problems due to the shaft's slow rotational speed (8.17 r.p.m.).

Solution

Temporary repairs were carried out and replacement gearboxes were ordered (which had a lead time of 44 weeks). The gearboxes continued to operate satisfactorily with the close monitoring provided by the CENT service until the new gearboxes were fitted.

Acknowledgement

Reproduced from McMahon, R. J. (1993). Condition monitoring at a continuous coal handling facility. *MPhil Thesis*, University of Northumbria.

11.5 Fault diagnosis of ceiling tiles production

Ceiling tiles are widely used in large commercial buildings. Although there are many kinds, most of them have a fibrous nature, are approximately 1 m square

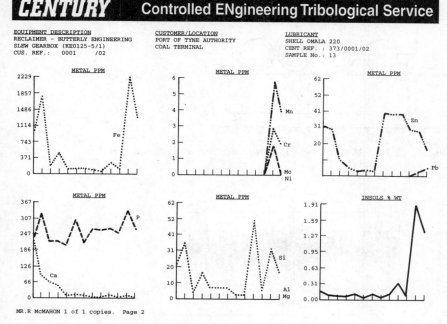

Figure 11.13   *CENT report: MORE*

and 20 mm thick. Their surface finish is produced by passing them through a sanding machine. A manufacturer had noticed that ripples sometimes ran across the tile surface, parallel to the sides and perpendicular to the direction of travel. Although the ripples had a small amplitude (50 μm), their appearance was obvious, especially near lights, so large numbers of tiles had to be scrapped. Vibration monitoring was employed to identify the cause of the problem and to discover a remedy.

Problem

The sanding machine consisted of four grinding drums approximately 1.2 m wide (Figure 11.14). The drums were driven by four electric motors which rotated at 1485 r.p.m. as the boards were fed through the machine at speeds of up to 0.7 m s^{-1}. Piezoelectric accelerometers were attached to the machine at the positions indicated; the signals were then played back through a dual-channel FFT analyser. Figure 11.15 shows the averaged spectrum from position 4 when the machine was producing poor-quality boards. The dominant peak is at the motor rotational frequency of 24.75 Hz and is due to unbalance of the drums.

However, the ripples on the board are due to variation in the vertical displacement of the drums relative to the surface of the board and bedplate. The displacement spectrum was obtained by double integration of the acceleration signal, and Figure 11.6 shows a dominant component at 5 Hz. This frequency

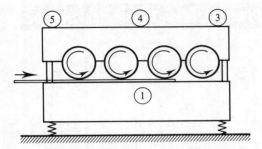

Figure 11.14   *Measurement locations on the sanding machine.*

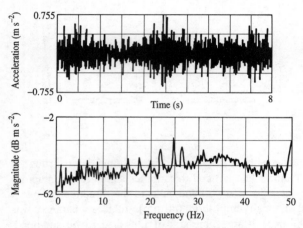

Figure 11.15   *Displacement spectrum for point 4.*

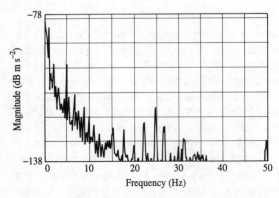

Figure 11.16   *The component at 5 Hz is dominant.*

correlated reasonably well with the velocity of the board and the wavelength of the ripples, approximately 50 mm. Further confirmation was obtained by using a low-pass filter set at 6 Hz. Figure 11.17 shows the time history for position 4 subtracted from the time history for position 1, indicating a relative amplitude of 45 μm (comparable with the ripple amplitude).

Solution

The 5 Hz motion was caused by the natural frequency of the machine on its elastomeric mounts. Performance was improved by paying careful attention to setting and clamping the top half of the machine; a satisfactory trace is shown in Figure 11.18.

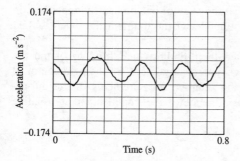

Figure 11.17   *Relative acceleration time history, position 1 – position 4: poor production quality.*

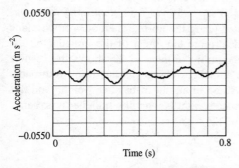

Figure 11.18   *Relative acceleration time history, position 1 – position 4: satisfactory production quality.*

11.6   Balancing problem

Bad unbalance may be made worse by badly fitting components; here is a straightforward case. It concerns small high-speed pumps which pump benzine through a large pressure rise in one stage. The pumps vibrated excessively with a frequency equal to the running speed immediately following commissioning. When a typical rotor and impeller (Figure 11.19) were dismantled and reassembled, the vibration levels were found to have changed but they remained constant for a given speed and load.

Approach

The user was a large petrochemical company staffed by competent engineers with a good knowledge of vibration. They quickly concluded that neither the rotor nor the rotor/impeller assembly had been adequately balanced, and the spline fit of the rotor/impeller assembly was so poor that the state of balance partly depended on the relative positions of the rotor and impeller at the moment they were bolted together.

The manufacturer initially rejected their findings and tried to blame the problems on faulty assembly of bearings, i.e. misalignment. So the petrochemical company's vibration engineers made the accurate measurements of the out-of-balance, particularly to determine the influence of the impeller and its relative angular position.

Results

The engineers' first action was to balance the rotors on their own to limits well inside the relevant values in BS 6861: Part 1: 1987. This small stiff rotor was treated like a rigid body and investigated using the normal methods of two-plane balancing. Their next action was to determine the effect of the impeller on the out-of-balance of the rotor/impeller assembly; Figure 11.20 shows the results. Altering the angular position of the impeller relative to the rotor is broadly consistent with altering the angular position of an out-of-balance weight; and this proved their conjecture, an out-of-balance weight is what the impeller represents.

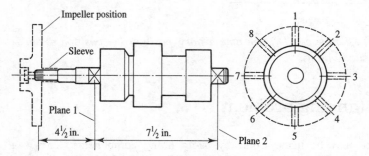

Figure 11.19   *Pump shaft (not to scale): all four are similar.*

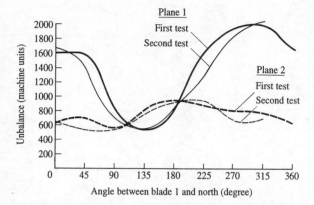

Figure 11.20   *Unbalance versus impeller engagement angle for shaft 3.*

Conclusion

The rotor/impeller assembly clearly needs to be balanced; this can be achieved within acceptable limits by removing mass from the impeller (the rotor having been balanced to very close limits), so that is what they did. For future stripdown and rebuild, the angular position of the impeller on the spline will be immaterial; this is a desirable state of affairs because during maintenance, it means there is one less thing to go wrong. And a closer fit in the spline was recommended so that a closely defined state of balance would be reproducible.

In the face of this evidence, the manufacturer admitted responsibility. This is a relatively straightforward case, which makes it all the more surprising that the manufacturer handled the rotational balance so badly and failed to appreciate the importance of properly fitting the spline.

11.7   Tool fly-cutter problem

This case study concerns a single-point tool fly-cutter running in air bearings at a speed of 9000 rev/min (Figure 11.21). When functioning normally it is capable of leaving a mirror finish with a single light cut on copper and aluminium alloys. This machine ran well for more than a year then had a period of producing poor surface finish, intermittently. The users were convinced that the problem was 'rotational balance', even though it could be very easily demonstrated that the rotor ran sweetly with very low levels of vibration normal to the axis of rotation.

The users were sustained in their belief that out-of-balance was the trouble by recalling a similar occurrence a few years earlier. On that occasion the problem seemed to disappear once the rotor had been balanced.

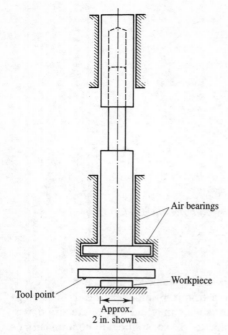

Figure 11.21   *Spindle assembly.*

Approach

By the time outside help was sought, the users were frantic over the disruption to their schedules. They were reluctant to believe the problem had any cause besides rotational balance and they could not conceive how their earlier 'cure' might have been spurious; the improvement following rotational balance could have been coincidental, not causal. That the problem took several years to return could simply be good luck in an unpredictable situation. But the users were eventually persuaded otherwise and they started to investigate the dynamic characteristics of the machine.

Results

The most difficult feature of this problem was the fact that it was intermittent. The machine still had periods when it cut perfectly well. When the machine performed badly, the erratic surface finish could be seen by inspection; aperiodic, it looked rather like the surface of a vinyl gramophone record. Indeed the surface finish was a recording of the tool-point motion.

The intermittent nature of vibration suggested that something was causing the system to flip between stable and unstable states. In one state it was capable of vibrating in a chaotic way and in the other it was not. Air bearings can be dynamically unstable in certain circumstances. Bearings experience irregular ingress of water and/or oil; and bearings experience natural fluctuations in the ambient temperature (the machine was not housed in a temperature-controlled enclosure). Could these two factors combine to take the bearings into and out of regions of dynamic instability?

Conclusion

Attention became focused on the chaotic nature of the vibrations, not the fact they were intermittent. It was recalled that certain mechanisms which simultaneously exhibit backlash and dry friction are capable of responding chaotically when excited by a sine wave. The tool-point was receiving a rectangular pulse once per revolution as it struck the workpiece, traversed its surface and then left (Figure 11.22). A Fourier spectrum of this sequence gave a large number of sinusoidal excitation components of different frequencies and amplitudes (Figure 11.23).

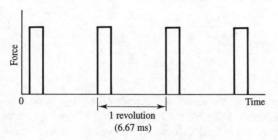

Figure 11.22 *General form of force components.*

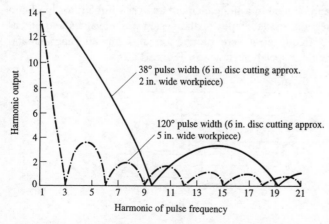

Figure 11.23 *Harmonic output for rectangular pulses.*

Although the investigation proved long and frustrating, it finally came up with an answer. On the umpteenth stripdown, it was discovered that *parts were loose*. The parts were found in the section of the table mechanism which carried the workpiece; inserts of the wrong size had been used so that vital components were separated by clearance instead of being nipped hard together as intended. This was a mechanism which simultaneously exhibited friction and backlash – it was capable of chaotic behaviour.

When all parts were assembled tightly, the trouble never recurred. The intermittent behaviour has not been satisfactorily explained. Perhaps the vibration ratcheted the grease into build-ups, producing a temporary increase in local solidity, temporarily cancelling the backlash; perhaps not.

11.8   Analysis of compressor vibration

A chemical plant compressor was found to suffer severe vibration problems at an operating speed of 4600 r.p.m. The vibration was so severe that it prevented the machine from being operated at its design speed of 5000 r.p.m. The vibration was found to be synchronous but could not be reduced by balancing. An analysis was undertaken.

Problem

The compressor was driven by a steam turbine and supported by two elliptical fluid film bearings. Figure 11.24 shows the basic dimensions. The bearings were analysed to determine the displacement and velocity coefficients over the operating range of the machine. Figure 11.25 shows the coefficients plotted against Sommerfeld number ($S = \eta NLDR^2/Wc^2$). The stiffness and damping properties of the bearings were then included in a finite element analysis of the rotor to obtain the natural frequencies and mode shapes of the system (eigenvalues and eigenvectors). The natural frequencies were plotted on a Campbell diagram and compared with the running speed as shown in Figure 11.26.

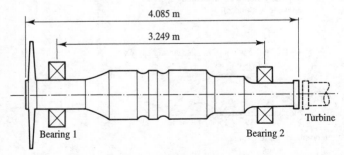

Figure 11.24   *Compressor shaft showing basic dimensions.*

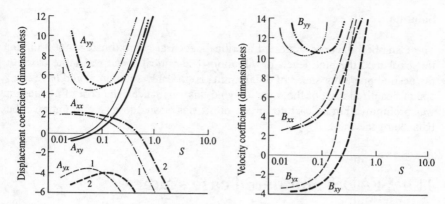

Figure 11.25 *(a) Displacement and (b) velocity coefficients for the two-lobed bearings with* $\delta = 0.5$: *(1) bearing 1,* $L/D = 1.2$; *(2) bearing 2,* $L/D = 0.73$.

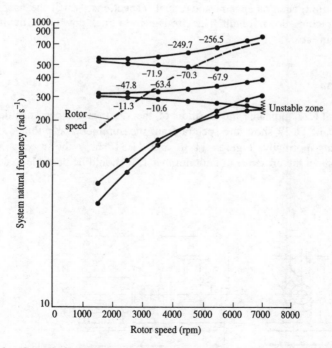

Figure 11.26 *Campbell diagram for the bearings: the number beside each point is the value of the real part of the eigenvalue at that point.*

Solution

The Campbell diagram indicated a critical speed at approximately 4600 r.p.m.; this explained the high levels of vibration. The critical speed was quite close to the desired operating speed of 5000 r.p.m., so the only solution would have been a significant redesign of the rotor to avoid this resonance. The cost of a redesign was prohibitive but at least no further effort was wasted in trying to improve the balance of the rotor.

11.9 FADS: simulated case studies

The University of Strathclyde has developed a computer-based system for fault analysis and diagnosis simulation (FADS); it enables engineers to gain experience of diagnosing faults in a typical system. The plant modelled here consists of a centrifugal pump driven at 2850 r.p.m. by an induction motor (Figure 11.27). Users of the system are provided with basic data and spectra taken from measurements at six points on the plant when operating satisfactorily and in good condition. Five faults have been modelled and spectra and inspection reports have been produced to reflect each fault. Costs are specified for obtaining further information, such as spectra and visual inspections, which the user can request when tracking down a fault. The aim is then to track down a particular fault with minimum cost.

Problem

Figure 11.27 indicates the location of the six measurement positions. Figures 11.28 and 11.29 show the spectra from measurements at points 1 and 4 when operating normally. Figures 11.30 and 11.31 show the spectra at the same locations in the presence of fault number 1. It should be possible to determine the

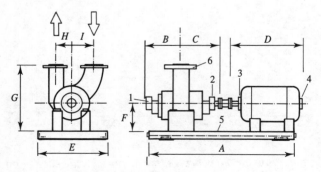

Figure 11.27 *Motor/pump unit.*

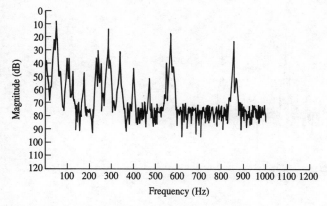

Figure 11.28 *Normal operation: spectrum at point 1.*

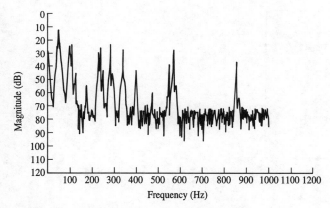

Figure 11.29 *Normal operation: spectrum at point 4.*

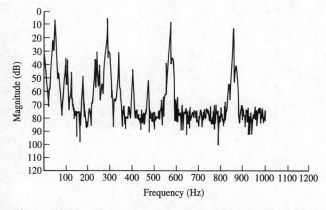

Figure 11.30 *Operation with fault 1: spectrum at point 1.*

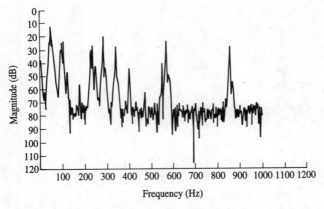

Figure 11.31   *Operation with fault 1: spectrum at point 4.*

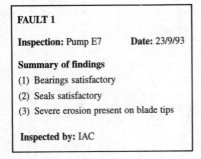

FAULT 1

Inspection: Pump E7   **Date:** 23/9/93

Summary of findings

(1) Bearings satisfactory

(2) Seals satisfactory

(3) Severe erosion present on blade tips

Inspected by: IAC

Figure 11.32   *Pump inspection report.*

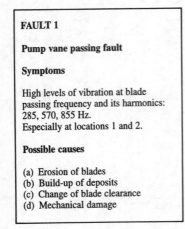

FAULT 1

Pump vane passing fault

Symptoms

High levels of vibration at blade
passing frequency and its harmonics:
285, 570, 855 Hz.
Especially at locations 1 and 2.

Possible causes

(a) Erosion of blades
(b) Build-up of deposits
(c) Change of blade clearance
(d) Mechanical damage

Figure 11.33   *Diagnosis sheet.*

likely fault from these spectra, but often the user will have requested much more data by the time these spectra are obtained, and by then may well be barking up the wrong tree. The inspection report for the pump should confirm the diagnosis (Figure 11.32) but this costs a lot more to obtain. When the user has correctly identified the problem, confirmation is provided along with a list of what should have been observed (Figure 11.33).

Conclusion

The FADS system provides realistic experience of the typical problems faced by maintenance staff. Cost penalties for an incorrect diagnosis are fictitious but the sense of achievement in correctly identifying the fault is real. The process shows users the immense benefit of hindsight! The simulation draws together the theory and practice of vibration measurement and fault diagnosis in an enjoyable and safe environment.

Bibliography

Introduction

The past decade has witnessed a very large number of publications on condition monitoring and related subjects. To supplement the further reading at the end of the chapters, we have compiled a bibliography from a literature search covering about ten years. We have tried to concentrate on publications that describe applications of condition monitoring, but where appropriate, we have listed theoretical studies, too. Apart from textbooks and journal publications, a great deal of the information was gleaned from three sources:

- The Engineering Index

- Applied Science and Technology Abstracts as supplied by H.W. Wilson & Co.

- COMADEM Conference Proceedings

The information is arranged in four major sections:

- General machinery

- Power plants and turbomachinery

- Bearings and gears

- Textbooks, standards and company handbooks

General machinery

Agrawal B. L., Demcko J. A. *et al.* (1992). Shaft torque monitoring using conventional digital fault recorders. *IEEE Trans. on Power Systems*, **7**, 1211–17.

Angelo M. (1990). Choosing accelerometers for machinery health monitoring. *Sound & Vibration*, **24**, 20–24.

Anvar A. M. and Kuhnell B. T. (1994). Comparison of classification parameter techniques using displacement domain signals for internal combustion engine fault diagnosis (part 1). In *Proc. COMADEM'94*, Sept.

Arreguy D. and Decarvalho S. S. (1990). A monitoring system for an NC machine tool. In *Proc. Signal Processing and System Control Factory: 16th Amer. Conf. IEEE, Industrial Electronics Society*.

Au-Yang M. K. (1993). Application of ultrasonics to non-intrusive vibration measurement. *J. Press. Vess. Tech.*, **115**, 415–19.

Azzam H. and Andrew M. J. (1992). The use of math-dynamic models to aid the development of integrated health and usage monitoring systems. *Proc. I. MechE, Pt. G*, **206**(1), 71–76.

Baguley P. (1986). Condition-based maintenance. *Chemical Engineer*, **427**, 38–40.

Baker R. D. (1991). Testing the efficacy of preventative maintenance. *J. Oper. Res. Soc.*, **42**(6), 493–503.

Barasch M. (1984). Modulation: a problem in vibration trouble-shooting. *PIMA. Mag.*, **66**, 51–55.

Bently D. E., Zimmer S. *et al.* (1986). Interpreting vibration information from rotating machinery. *Sound & Vibration*, **20**, 14–23.

Berggren C. (1994). Reaping the benefits of online predictive maintenance. *Ind. Eng.*, **26**, 18–19.

Berry J. E. (1986). Diagnostic evaluation of machinery using vibration signature analysis. *Sound & Vibration*, **20**, 10–17.

Berry J. E. (1991). How to specify machinery vibration spectral alarm bands. *Sound & Vibration*, **24**(9), 16–26.

Bogard W. T., Palusamy S. S. *et al.* (1988). Applying automation to diagnostics, predictive maintenance in plants. *Power*, **132**, 27–28.

Boggs R. E. (1990). Rotating machine misalignment: the silent disease but not the symptom. *Tappi J.*, **73**, 111–16.

Braun S. (1983). MSA: mechanical signature analysis. In *Proc. Conf. on Mechanical Vibration and Noise,* Michigan, Sept.

Braun S. (1991). Vibration-based diagnostics. *Cond. Mont. Diag. Tech.*, **1**(4), 136–43.

Brown D. N. (1991). Envelope and cepstrum analysis: two vibration analysis techniques helping to maintain production output and paper quality. *Tappi J.*, **74**, 5–8.

Brown R. and Chatingny J. V. (1988). Improved transducers for vibration monitoring. *Machine Design*, **60**, 135–36.

Burrows J. H. (1991). Optimising the use of periodically collected data for machine condition monitoring. In *Proc. 3rd. Int. Cong. on Condition Monitoring and Diagnostic Engineering*, **60**, 135–36.

Caribbio B. (1989). Automating signature analysis. *Automation*, **36**, 18.

Chandler J. (1991). Meaningful vibration measurements for predictive maintenance. *Sound & Vibration*, **25**, 18–22.

Chilcott J. B. and Christer A. H. (1991). Modelling of condition-based maintenance at the coalface. *Int. J. Prod. Econ.*, **22**, 1–11.

Chin H. and Danai K. (1991). A method of fault signature extraction for improved diagnosis. *J. Dyn. Sys. Meas. Cont.*, **113**, 634–38.

Cory W. T. M. (1991). Overview of condition monitoring methods with emphasis on industrial fans. *Proc. I. MechE. Pt. A*, **205**(4), 225–40.

Craighead I. A. and Thomas M. R. (1994). FADS – a vibration monitoring/fault diagnosis game. In *Proc. COMADEM'94*, Sept.

Cue R. W. and Muir D. E. (1991). Engine performance monitoring and troubleshooting for the CF-18 aircraft. *J. Eng. for Gas Turbines & Power*, **113**, 11–19.

Culp C. H. (1989). Expert systems in preventative maintenance and diagnostics. *ASHRE J.*, **31**, 24–27.

De Chiffrel L. (1984). Frequency analysis of surfaces machined using different lubricants. *ASLE Trans.*, **27**, 220–26.

Dohner C. (1988). Expert system for gas turbines. *EPRI J.*, **13**, 44–45.

Dundics M. J. and Erickson D. P. (1989). Machinery maintenance enters the digital world. In *Proc. ShipTech. Res. STAR Symp. 21st Century Ship and Offshore Vessels*.

Elkordy M. F. and Chang K. C. (1994). Application of neural networks in vibration signature analysis. *J. Eng. Mech.*, **120**(2), 250–65.

Emel E. *et al.* (1991). Acoustic emission monitoring of the cutting process. *J. Eng. Mater. Tech.*, **113**, 456–64.

English J. R., Foster J. W. and Bates T. E. (1990). Developing techniques for achieving reliability in purchased equipment. In *Proc. Int. Industrial Engineering Conf.*

Fahmy M. N. *et al.* (1984). Cepstrum analysis of surface waves in acoustic signature inspection of railroad wheels. *J. Acoust. Soc. Amer.*, **75**, 1283–90.

Finkel C. M. (1993). Legal preventative maintenance. *Air Progress*, **55**, 22–23.

Follmann D. A. and Goldberg M. S. (1988). Distinguishing heterogeneity from decreasing hazard rates. *Technometrics*, **30**, 389–96.

Frarey J. L. and Eisler, E. A. (1986). Remote monitoring of machinery. *Sound & Vibration*, **20**, 14–18.

Gasparic J. J. (1991). Vibrational analysis identifies the causes of mill chatter. *Iron & Steel Eng.*, **68**, 27–29.

Gilstrap M. (1984). Transducer selection for vibration monitoring of rotating machinery. *Sound & Vibration*, **18**, 22–24.

Glaskin M. (1990). Monitoring progress. *Engineering(London)*, **230**, 21.

Gorter J. and Klijn A. J. (1986). Vibration measurements provide condition monitoring of rotating equipment. *Oil & Gas J.*, **84**, 64–70.

Gu F. and Ball A. D. (1994a). Transient signal analysis using the Wigner Ville distribution. In *Proc. COMADEM'94*, Sept.

Gu F., Ball A. D. *et al.* (1994b) Identification of diesel fuel injection characteristics for the time–frequency analysis of vibration signatures. In *Proc. COMADEM'94*, Sept.

Gustafson D. A. (1990). New age in engine monitoring. *Air Progress*, **52**, 58–61.

Hafmann S. L. (1987). Vibration analysis for preventative maintenance: a classical case history. *Marine Tech.*, **24**, 329–39.

Halfen E. M. (1985). Predictive maintenance can save you money. *InTech*, **32**, 111–13.

Haran S., Finch R. D. *et al.* (1989). Application of an automated package of pattern recognition techniques to acoustic signature inspection of railroad wheels. *J. Acoustic. Soc. Amer.*, **85**, 440–49.

Hartley W. A. and Flanders E. E. (1985). A planned approach to machinery vibration and monitoring diagnostics. *Tappi J.*, **68**, 78–82.

Heckman A. T. (1993). Vibration monitoring yields big benefits. *Chemical Engineering*, **100**, 126, 129.

Hickling R. and Marin S. P. (1988). Enhancement of the sound power of a component of a complex noise source by sound from other nearby components. *J. Acoust. Soc. Amer.*, **84**, 262–74.

Hill E. and Lewis T. J. (1985). Acoustic emission monitoring of a filament-wound composite rocket-motor case during hydroproof. *Mater. Eval.*, **43**, 859–63.

Himelblau H. and Piersol A. G. (1989). Evaluation of a procedure for the analysis of non-stationary vibroacoustic data. *J. Environ. Sci.*, **32**, 35–42.

Hoch R. R. (1990). Practical application of reliability-centred maintenance. *ASME Paper, Jt. ASME/IEEE Power Generation Conf.*

Hodgson D. C. and Sadek M. M. (1983). A technique for the prediction of the noise field from an arbitrary vibrating machine. *Proc. IMechE., Pt. C*, **197**, 189–97.

Hoffner J., Olenick J. E. *et al.* (1991). Predictive maintenance for no-twist rod mills using vibration signature analysis. *Iron & Steel Eng.*, **68**, 55–61.

Holmberg K., Enwald P. *et al.* (1994). Maintenance and reliability: advanced technologies and technological trends. In *Proc. COMADEM'94*, Sept.

Holroyd T. J., Kings S. D. *et al.* (1990). Machine condition monitoring via stress wave sensing. *IMechE. Publ. on Machine Condition Monitoring*, 45–48.

Holyrod T. J. and Randall N. (1993a) Use of acoustic emission for machine condition monitoring. *Br. J. NDT*, **35**(2), 75–83.

Holyrod T. J. and Randall N. (1993b). Field application of acoustic emission to machinery condition monitoring. In *Proc. 5th. Int. Cong. on Condition Monitoring and Diagnostic Engineering*.

Holyrod T. J. and Randall N. (1994c). Applying acoustic emission to industrial machinery monitoring. In *Condition Monitoring '94*.

Holyrod T. J. and Randall N. (1994d). Cost-effective condition monitoring based on acoustic emission. In *Proc. Ann. Conf. BINDT*, Sept.

Hong S. Y., Ni J. *et al.* (1994). Pre-emptive diagnosis of minor machine failure by DDS spectrum analysis. *J. Eng. for Industry*, **116**, 130–33.

Janisz C. K. and Brokenshire R. E. (1985). Diagnosing machinery problems with automated vibration analysis. *Sound & Vibration*, **19**, 14–17.

Jones R. M. (1989). Acoustic monitoring of nuclear plant operating valves for leakage. *Mater. Eval.*, **47**, 1278.

Kasai S., Tada K. and Hasegawa T. (1991). Development of Diagnosis Techniques for Rotating Machinery by Vibration Analysis. *Kawasaki Steel Technical Report 24*, 54–63.

Kasyanov V. E. (1990). Integral assessment, improvement and optimisation of the reliability of machines. *Sov. Eng. Res.*, **10**(4), 1–3.

Katzel J. (1987). Applying predictive maintenance. *Plant Eng.*, **41**, 62–65.

Kim B. S. (1983). Punch press monitoring with acoustic emission. *J. Eng. Mater. Tech.* **105**, 295–306.

Kim H. M. and Bartkowick T. J. (1993). Damage detection and health monitoring of large space structures. *Sound & Vibration*, **27**, 12–17.

Korpi K. and Ahlbom K. (1986). Online condition monitoring and lubrication system for paper mill. *Tappi J.*, **69**, 173–75.

Kruger W. L. (1989). Digital protection systems provide more than monitoring for rotating machinery. *Control Eng.*, **36**, 131–32.

Kumar A. (1991). Condition-based predictive maintenance. In *Proc. 3rd. Int. Cong. on Condition Monitoring and Diagnostic Engineering, Etc.*

Lacoste M. and Rogard V. (1988). Mediatised interaction between experts in the maintenance of automated machines. In *IFAC Proc. Ser. No. 3: Analysis, Design and Evaluation of Man–Machine Systems*.

Laws W. C. and Muszynska A. (1987). Periodic and continuous vibration monitoring for preventative/predictive maintenance of rotating machinery. *J. Eng. for Gas Turbines & Power*, **109**, 159–67.

Leon R. L. (1985). Is your periodic machinery monitoring program telling you the truth?. *Sound & Vibration*, **19**, 24–26.

Lewis W. S. (1991). Vibration analysis is the cornerstone of predictive and preventative maintenance. *Tappi J.*, **74**, 77–82.

Liddle I. and Reilly S. (1993). Automatic analysis of rotating machinery using an expert system. *Sound & Vibration*, **27**, 6.

Lifson A., Quentin G. H. and Smalley A. J. (1989). Assessment of gas turbine vibration monitoring. *J. Eng. for Gas Turbines & Power*, **111**, 257–63.

Lifson A., Harold R. and Smalley A. J. (1987). Vibration limits for rotating machinery. *Mech. Eng.*, **109**, 60–63.

Lin C. E., Huang C. J. and Huang C. L. (1992). An expert system for generator maintenance scheduling using operation index. *IEEE Trans. on Power Systems*, **7**, 1141–48.

Lin C. C. and Wang H. P. (1993). Classification of autoregressive spectral estimated spectral patterns using an adaptive resonance theory neural network. *Computers in Industry*, **22**, 143–57.

Litz J. (1984). High-powered analysis of vibration spectra. *Machine Design*, **56**, 79–83.

Lodge C. (1987). Predictive systems cut repair costs by 50%. *Plastics World*, **45**, 70–71.

Lopatinskaia E., Zhu J. *et al.* (1994). Monitoring the vibrations of varying-speed machinery using a recursive filter and the angle domain. In *Proc. COMADEM'94*.

Luo M. F. and Mathew J. (1993). Monitoring machinery with varying speed using angle-domain analysis. *In Proc. 17th. Meeting on Vibration Instruments*, St. Louis MO.

Lyon R. H. (1988). Vibration-based diagnostics of machine transients. *Sound & Vibration*, **22**, 18–22.

McGuckin W. J. and Schramm E. J. (1985). Diagnostic analysis of machinery with state-of-the-art equipment. *Sound & Vibration*, **19**, 6–8.

Maxwell H. (1985). Base your vibration monitoring plan on maintenance savings. *Power*, **129**, 35–37.

Meeker J. (1985). Applying condition monitoring equipment: a case study. *Process Eng.*, **66**, 71.

Mikulski R. (1988). Limerick nuclear generating station vibration monitoring system. *IEEE Trans. on Energy Conversion*, **3**, 531–35.

Miller G. P. and McClymonds S. L. (1990). Maintenance cost avoidance through comprehensive condition monitoring. *ASME Paper, Jt. ASME/IEEE Power Generation Conf.*

Miller J. (1991). How to conduct predictive machinery vibration programs. *Tappi J.*, **74**, 77–82.

Mitchell J. S. (1985). Condition monitoring. *Mech. Eng.*, **107**, 32–37.

Muhammad A., King G. A. *et al.* (1994). Genetic algorithms and digital signal processing in condition monitoring. In *Proc. COMADEM'94*, Sept.

Mullen M. and Richter B. (1993). Integrated diagnostics for aircraft propulsion systems. In *Proc. IEEE Systems Readiness Tech. Conf.* San Antonio TX, Sept.

Murphy B. R. and Watanabe I. (1992). Digital shaping filters for reducing machine vibration. *IEEE Trans. on Robotics & Automation*, **8**, 285–89.

Narayanan T. S. (1993). Eeffective online monitoring: the role of potential time measurements. *Metal Finishing*, **91**, 57–59.

Nilsson E. (1992). The burden of preventative maintenance, a manufacturer's challenge. *CIM Bull.*, 52–54.

Nordwall B. D. (1992). Engine monitoring system may provide 100 hour warning before jet failure. *Aviation Weekly & Space Tech.*, **136**, 54–55.

Page, E. A. and Berggren C. (1994). Improve predictive maintenance with HFE monitoring. *Hydrocarbon Processing*, **73**, 69–72.

Pocock S. and Allen J. (1986). Machine condition monitoring – are you convinced yet? *Process Eng.*, **67**, 41.

Poon H. L. (1990). Verbal time series reports generation in condition monitoring. *Computers in Industry*, **15**(4), 293–301.

Potter R. (1990). A new order tracking method for rotating machinery. *Sound & Vibration*, **24**, 30–34.

Powell C. D. (1987). Machinery analysis and monitoring. *Sound & Vibration*, **21**, 18.

Powell C. D. and Dietz C. P. (1985). A vibration control case history trilogy. *Sound & Vibration*, **19**, 18–23.

Powers B. L. and Chuang S. Y. (1991). Vibration monitoring of small lightweight components using fibre-optic sensors. *J. Testing and Evaluation*, **19**, 493–96.

Qu L. and Xu G. (1991). Investigation of a special trending method for rotating machinery monitoring. *Cond. Mont. Diag. Tech.*, **1**(3), 84–88.

Quintas A. C. (1984). Monitoring equipment wear in a paper and pulp plant. *Lubrication Eng.*, **40**, 648–58.

Randorff J. E., Spillman R. R. and Rolen T. J. (1989). Machinery protection through vibration monitoring. *Public Works*, **120**, 58–60.

Rao P., Taylor F. *et al.* (1990). Real-time monitoring of vibration using the Wigner distribution. *Sound & Vibration*, **24**, 22–25.

Ray A. K. (1991). Equipment fault diagnosis: a neural network approach. *Computers in Industry*, **16**, 169–77.

Reason J. (1985). Vibration monitoring: systematic data collection/analysis is key to higher availability. *Power* **129**, 1–12.

Reason J. (1987). Continuous vibration monitoring moves into diagnostics. *Power*, **131**, 47–51.

Rees F. (1987). Monitoring system to predict machine failure. *Information & Software Tech.*, **29**, 6–7.

Renwick J. T. (1984). Condition monitoring of machinery using computerised vibration signature analysis. *IEEE Trans. on Industry Applications*, **20**, 519–27.

Renwick J. T. and Babson P. E. (1985). Vibration analysis: a proven technique as a predictive maintenance tool. *IEEE Trans. on Industry Applications*, **21**, 324–332.

Reynolds G. and Coley T. (1994). Boise cascade uses machinery monitoring system to avoid catastrophic failures. *Sound & Vibration*, **28**, 6–7.

Robinson R. C. (1993). Corrosion monitoring augmented by waveform analysis. *Pipe Line Industry*, **76**, 21–24.

Sachs N. W. (1991). What does PM really mean? *Lubrication Eng.*, **47**, 889–91.

Salzberg A. A. (1989). Validation of LACE spacecraft vibroacoustic prediction model. *J. Environ. Sci.*, **32**, 53–59.

Serridge M. (1989). Fault detection techniques for reliable condition monitoring. *Sound & Vibration*, **23**, 18–22.

Serridge M. (1990). Better machinery monitoring. *Hydrocarbon Processing*, **69**, 53–56.

Shreve D. H. (1990). Machinery condition surveillance systems. *Adv. Instrument. Proc., Pt. 4*, **45**, 1737–44.

Smiley R. G. (1983). Rotating machinery: monitoring fault diagnosis. *Sound & Vibration*, **17**, 26–28.

Smiley R. G. (1984). Expert system for machinery fault diagnosis. *Sound & Vibration*, **18**, 26–28.

Smith J. D. (1987). Diagnostic analysis leads the way in preventative maintenance. *Power Eng.*, **91**, 12–19.

Soffker D. and Muller P. C. (1993). Crack detection in turbo rotors: vibration analysis and fault detection. In *Proc. 14th ASME Design Conf. on Noise and Vibration*, Albuquerque NM, Sept.

Staszewski W. J. and Worden K. (1993). Classification of faults in spur gears. In *Proc. World Cong. on Neural Networks*, Portland OR.

Steven R R. (1984). Preventative maintenance depends on planning. *Offshore*, **44**, 48.

Stewart R. M. (1985). A systematic approach to automating machinery management. *Sound & Vibration*, **19**, 14–18.

Strom U. (1989). Process and condition monitoring can solve the quality/speed dilemma. *PIMA Mag.*, **71**, 56–57.

Tan C. C. (1990). Transducer systems for condition monitoring using adaptive noise cancellation. *Natl. Conf. Publ. Inst. Aust.*, No. 90, Pt. 14.

Taylor J. I. (1985). Diagnosing paper machine problems with vibration analysis. *PIMA Mag.*, **67**, 61–62.

Taylor J. L. (1988). Accurate predictive maintenance programs. *Tappi J.*, **71**, 113–17.

Taylor J. I. (1992). State-of-the-art predictive maintenance programs. *Tappi J.*, **75**, 75–78.

Teramae K. (1991). Portable Machine Analyser for Equipment Diagnosis. *Kawasaki Steel Technical Report* 24, 109–11.

Thibault S. E. *et al.* (1994). NTD system prevents catastrophic failures. *Power Eng.*, **98**, 37–40.

Thomas M. and Foster C. (1994). Developing an implementation strategy for condition monitoring. In *Proc. COMADEM'94*, Sept.

Thomas V. C. (1987). A vibroacoustic database management centre for shuttle and expendable launch vehicle payloads. *Journal of Environmental Sciences*, **30**, 24–26.

Tirinda P. and Ballo I. (1986). Some problems of the computerised system for multiple machinery survey and the online permanent monitoring of large centrifugal compressor operation. *Computers in Industry*, **7**, 15–22.

Tranter J. T. (1990). The application of computers to machinery predictive maintenance. *Sound & Vibration*, **24**, 14–19.

Tranter J. T. (1990). Fundamentals of and application of computers to condition monitoring and predictive maintenance. *Natl. Conf. Publ. Inst. Eng. Aust.*, No. 90, Pt. 9.

Turley W. (1992). Vibration analysis system can cut maintenance costs. *Rock Products*, **95**, 21.

Vaija P. *et al.* (1985). Multilevel failure detection system. *Computers in Industry*, **6**, 253–63.

Wada K. (1991). Online Machine Diagnosis System. *Kawasaki Steel Technical Report* 24, 112–14.

Watts W. *et al.* (1993). Automated vibration-based expert diagnostic system. *Sound & Vibration*, **27**(9), 14–20.

White G. (1991). Amplitude demodulation: a new tool for predictive maintenance. *Sound & Vibration*, **25**, 14–19.

Whittington H. W. and Flynn B. W. (1993). High-reliability condition monitoring systems. *Br. J. NDT*, **35**(11), 648–54.

Withers W. D. and Davis J. (1990). Increasing paper machine speeds with diagnostic vibration analysis and on-site balancing and repairs. *Tappi J.*, **73**, 33–36.

Xu Yin-Ge (1991). Research on the Haar spectrum in fault diagnosis of rotating machinery. *Appl. Math. Mech. Eng. Educ.*, **12**(1), 61–66.

Yang R., Bawden W. F. *et al.* (1993). An integrated technique for vibration monitoring adjacent to a blast hole. *CIM Bul.*, **86**, 45–52.

Yao G. C., Chang K. C. *et al.* (1992). Damage diagnosis of steel frames using vibrational signature analysis. *J. Eng. Mech.*, **118**, 1949–61.

Yeung K. K. *et al.* (1994). Development of computer-aided image analysis for filter debris analysis. *Lubrication Eng.*, **50**, 293–99.

Zaharko R. L. (1988). History of predictive maintenance at St. James River's Wauna Mill. *Tappi J.*, **71**, 71–72.

Zhang B. Q. (1989). Acoustic emission online monitoring for petrochemical plants. *Mater. Eval.*, **47**, 351–55.

Zheng X, and Yang S. (1989). Plant condition recognition: a time series model approach. *Computers in Industry*, **11**, 333–40.

Zhengjia H., Zhao J. *et al.* (1994). Wavelet transform in tandem with autoregressive technique for monitoring and diagnosis of machinery. In *Proc. COMADEM'94*, Sept.

Power plants and turbomachinery

Akbari H. *et al.* (1988). Use of energy management systems for performance monitoring of industrial load-shaping measures. *Energy*, **13**, 253–63.

Applegate P. (1990). Balanced forces and foundations: keys to reciprocating compressor life. *Oil & Gas J.*, **88**, 50.

Avrunin G. A. (1989). On choosing the conditions for the diagnosis of the technical state of hydraulic motors. *Sov. Eng. Res.*, **9**, 37–39.

Baur P. (1983). Field balancing of rotating machinery. *Power*, **127**, 1–16.

Berzonsky B. E. (1990). A knowledge-based electrical diagnostic system for mining machine maintenance. *IEEE Trans. on Industry Applications*, **26**, 342–46.

Bigret R., Coetzee C. J. *et al.* (1986). Measuring the torsional modal frequencies of a 900 MW turbogenerator. *IEEE Trans. on Energy Conversion*, **1**, 99–105.

Boness R. J. and McBride S. L. (1991). Condition monitoring of adhesive and abrasive wear processes using acoustic emission techniques. In *Proc. 3rd. Int. Cong. on Condition Monitoring and Diagnostic Engineering*.

Braig R., Paris J. *et al.* (1985). Dynamic balancing of gas-lubricated high-speed turbines. *Cryogenics*, **25**, 638–40.

Branagan L. (1991). Data interpolation for vibration diagnostics using two-variable correlations. In *Recent Innovations and Experience with Plant Monitoring and Utility Operation*, Vol. 15. ASME.

Calandranis J. *et al.* (1990). DIAD–Kit/Boiler: online performance monitoring and diagnosis. *Chem. Eng. Prog.*, **86**, 60–68.

Cameron J. R., Thomson W. T. and Dow A. B. (1986). Vibration and current monitoring for detecting air-gap eccentricity in large induction motors. *Proc. IEE, Pt. B*, **133**, 155–63.

Cartwright R. A. and Fisher, C. (1991). Marine gas turbine condition monitoring by gas path electrostatic detection technique. *ASME Paper, Int. Gas Turbine and Aero-engine Cong.*

Chilcott J. B. and Christer A. H. (1991). Modelling of condition-based maintenance at the coalface. *Int. J. of Prod. Econ.*, **22**, 1–11.

Chudnoff R. (1987). Monitoring the performance of power plants. *Mech. Eng.*, **109**, 56–58.

Conforti G. *et al.* (1989). Fibre-optic vibration sensor for remote monitoring in high-power electric machines. *Appl. Opt.*, **28**, 5158–61.

Cory W. T. W. (1991). Overview of condition monitoring methods with emphasis on industrial fans. *Proc. IMechE., Pt. A*, **205**(4), 225–40.

Cosgrove J. A., Vourdas A. and Jones G. R. (1993). Acoustic monitoring of partial discharges in gas-insulated substations, using optical sensors. *Proc. IEE, Pt. A*, **140**, 369–74.

Cudworth C. J. and Smith J. R. (1990). Steam turbine generator shaft torque transients: a comparison of simulated and test results. *Proc. IEE, Pt. C*, **137**, 327–34.

Cullen J. P. (1988). Monitoring system improves maintenance for North Sea industrial gas turbines. *Oil & Gas J.*, **86**, 71–75.

DeMoss S. H. (1991). Combined gas turbine control and condition monitoring. *ASME Paper, Int. Gas Turbine and Aero-engine Cong.*

Dennis S. J. (1990). Condition monitoring in the food and power industry. *Natl. Conf. Publ. Inst. Eng. Aust.*, No. 90, Pt. 9.

Dimapogonas E. D. (1993). Development of learning expert systems for early diagnostics of turbine machines on the basis of neural networks. *Teploenerg*, No. 10, 68–70.

Dodd V. R. (1984). Condition monitoring of major turbomachinery cuts costs over a 4 year period. *Oil & Gas J.*, **82**, 96.

Driedger W. C. (1990). Getting the picture with compressor monitoring. *InTech*, **37**, 46–47.

Edris A. A. (1993). Subsynchronous resonance countermeasure using phase imbalance. *IEEE Trans. on Power Systems*, **8**, 1438–44.

Eustace R. W. *et al.* (1994). Fault signatures obtained from fault implant tests on an F404 Engine. *ASME J. Eng. Gas Turbines & Power Trans.*, **116**(1), 178–83.

Fenton R. E., Gott B. E. B. *et al.* (1992). Preventative maintenance of turbine generator stator windings. *IEEE Trans. on Energy Conversion.* **7**, 216–22.

Gemmell B. D., McDonald J. R., Stewart R. W. and Weir B. J. (1991). An online monitoring system: embedded expert system for condition monitoring and fault diagnosis on turbo-alternators. In *Proc. 3rd Int. Cong. on Condition Monitoring and Diagnostic Engineering.*

Godse A. G. (1991). Implement machinery predictive maintenance. *Hydrocarbon Processing*, **70**, 163.

Goldman S. (1987). Vibration analysis now works on reciprocating engines. *Power*, **131**, 49–50.

Gonzalez A. J., Osborne R. L. *et al.* (1986). Online diagnosis of turbine generators using artificial intelligence. *IEEE Trans. on Energy Conversion*, **1**, 68–74.

Gough R. J. and Al-Shemmeri T. T. (1991). Strategy for monitoring, selective acquisition, data reduction and analysis system for the appraisal of compact heat exchangers. In *Proc. 3rd. Int. Cong. on Condition Monitoring and Diagnostic Engineering.*

Griffin J. H. (1992). Optimising instrumentation when measuring jet-engine blade vibration. *J. Eng. for Gas Turbine & Power*, **114**, 217–21.

Haddad S. D. and Cempel C. (1988). Optimising engine operation by monitoring the working gas pulsations. *J. Eng. for Gas Turbines*, **110**, 321–24.

Hammons T. J. (1989). Impact of shaft torsionals in steam turbine control. *IEEE Trans. on Energy Conversion*, **4**, 143–49.

Hammons T. J. and McGill J. F. (1993). Comparison of turbine generator shaft torsional response predicted by frequency domain and time methods following worst-case supply system events. *IEEE Trans. on Energy Conversion*, **8**, 559–65.

Harmon J. M., Napoli J. and Snyder G. (1992). Real-time diagnostics improve power plant operations. *Power Eng.*, **96**, 45–46.

Hess D. P., Park S. Y. *et al.* (1992). Non-invasive condition assessment and event timing for power circuit-breakers. *IEEE Trans. on Power Delivery*, **7**, 353–60.

Hoffner J., Olenick J. E. and Foley J. D. (1991). Predictive maintenance for no-twist rod mills using vibration signature analysis. *Iron & Steel Eng.*, **68**, 55–61.

Humes B. R. (1990). Vector monitoring pinpoints vibration in gas turbines. *Power*, **133**, 25–27.

Jennings G. D. and Harley R. G. (1990). Torsional interaction between non-identical turbine generators. *IEEE Trans. on Power Systems*, **5**, 133–39.

Kameda T. *et al.* (1993). Notes on multiple-input signature analysis. *IEEE Trans. on Computers*, **42**, 228–34.

Koch C. G., Isle B. A. *et al.* (1988). Intelligent user interface for expert systems applied to power plant maintenance and troubleshooting. *IEEE Trans. on Energy Conversion*, **3**, 71–77.

Kopczynski, W. (1985). Vibration analysis is hub of predictive maintenance scheme. *Water/Eng. & Management*, **132**, 36–37.

Kryter R. C. and Haynes H. D. (1989). Condition monitoring of machinery using motor current signature analysis. *Sound & Vibration*, **23**, 14–21.

Kurihara N., Nishikawa M. *et al.* (1984). A microprocessor-based vibration diagnostic system for steam turbines and turbogenerators in power plants. *IEEE Trans. on Power Apparatus & Systems*, **103**, 1283–91.

Lacey S. J. (1990). Vibration monitoring of the internal centreless grinding process. *Proc. IMechE., Pt. B*, **204**, 119–42.

Li Zuyun, Wang Xifeng, Shi Jianzhong and Wang Zheng (1991). Vibration fault diagnosis of boiler feed pumps. In *Proc. 3rd. Cong. on Condition Monitoring and Diagnostic Engineering*.

Lifson A. *et al.* (1989). Assessment of gas turbine vibration monitoring. *J. Eng. for Gas Turbines & Power*, **111**, 257–63.

Lloyd S. A. *et al.* (1992). Conditioning monitoring of ethylene main pipeline pumps. *Proc. IMechE.*, Pt. E, **206**(2), 93–97.

Loukis E. (1994). Optimising automated gas turbine fault detection using statistical pattern recognition. *ASME J. of Eng. Gas Turbines & Power*, **116**(1), 165–71.

Lucas H. (1988). Introduction and application of General Electric turbine engine monitoring software. *J. Eng. for Gas Turbines & Power*, **110**, 23–27.

McHattie L. and El-Alfy S. E. (1992). Maintenance improvements through the introduction of reliability concepts. *CIM Bull.*, **85**, 47–51.

McMahon R. J. and Craighead I. A. (1991). Condition monitoring at a continuous coal-handling facility. In *Proc. 3rd. Int. Cong. on Condition Monitoring and Diagnostic Engineering*.

Majovsky B. and Salamone D. J. (1988). Dynamic analysis of a steam turbine vibration problem. *Sound & Vibration*, **22**, 24–30.

Maxwell H. (1985). Base your vibration monitoring plan on maintenance savings. *Power*, **129**, 35–37.

Milenkovic V. *et al.* (1984). Online diagnostics of rear-axle transmission errors. *J. Eng. for Industry*, **106**, 331–38.

Mirro J. (1991). Turbomachinery vibration case histories: design problems of rotating machinery in operation. *Proc. IMechE.*, Pt. A, **205**(3), 183–93.

Morgan G. and Watton J. (1990). Online multi-gearbox monitoring of a wire-rod finishing mill. *Proc. IMechE.*, Pt. E, **204**(2), 103–9.

Muir D. E. *et al.* (1989). Health monitoring of variable-geometry gas turbines for the Canadian Navy. *J. of Eng. for Gas Turbines & Power*, **111**, 244–50.

Nieb J. R. and Nicolas V. T. (1991). Automated monitoring and control of vibration and chatter in rolling processes. *Iron & Steel Eng.*, **68**, 33–42.

Parascos E. T. (1992). Reliability, availability and maintainability in the electrical power industry. *Proc. American Power Conf.*, Pt. 1, **54**, 589–93.

Pardue E. F., Piety K. R. and Moore R. (1992). Elements of reliability-based machinery maintenance. *Sound & Vibration*, **26**, 14–20.

Piety K. R. and Pardue E. F. (1986). Predictive maintenance programs for the power generation industry. *Sound & Vibration*, **20**, 18–21.

Poley J. (1991). Oil analysis power user bringing the evaluation process in-house. *Lubrication Eng.*, **46**(10), 630–35.

Poon H. L. (1991). A knowledge-based condition monitoring system for electrical machines. *Computers in Industry*, **16**, 159–67.

Prouty R. W. (1988). Helicopter vibration. *Sound & Vibration*, **22**, 34–36.

Ricca P. M. and Bradshaw P. M. (1984). TAPS machinery monitoring program. *Oil & Gas J.*, **82**, 96–98.

Richards S. J. (1988). Motor vibration analysis: key to effective troubleshooting. *Power*, **132**, 46–48.

Roberton R. S. (1986). ASTM in-service monitoring program for steam and gas turbine oils. *Lubrication Eng.*, **42**, 466–73.

Schilling M. T., Praca J. C. G. and DeQueiroz J. F. (1988). Detection of ageing in the reliability analysis of thermal generators. *IEEE Trans. on Power Systems*, **3**, 490–97.

Schomer P. D. and Neathammer R. D. (1987). The role of helicopter noise induced vibration and rattle in human response. *J. Acoust. Soc. Amer.*, **81**, 966–76.

Schuster S. A. and Jurgeleit W. (1994). Real-time signal processing for a jet-engine vibration test system. *Sound & Vibration*, **28**, 18–21.

Scott W. B. (1983). USAF tests turbine monitoring system. *Aviation Weekly & Space Tech.*, **119**, 77.

Short J. D. and Durham M. O. (1988). Use reliability analysis to increase ESP run-life. *World Oil*, **206**, 49–51.

Stein J .L. and Shin K. (1986). Current monitoring of field-controlled DC spindle drives. *J. Dyn. Sys., Meas. Cont.*, **108**, 289–95.

Storey P. A. (1984). Holographic vibration measurement of a rotating flutter fan. *AIAA J.* **22**, 234–41.

Tack A J., Aplin P. F. and Cane B. J. (1991). Advanced condition monitoring for high-temperature applications. In *Proc. 3rd. Int. Cong. on Condition Monitoring and Diagnostic Engineering*.

Taylor J. I. (1985). Using spectrum analysis to identify gear problems. *PIMA Mag.*, **67**, 40–43.

Toler D. F and Yorio R. N. (1984). Operational mode monitoring of gas turbines in an offshore gas application. *J. Eng. for Gas Turbines & Power*, **106**, 940–45.

Turner M. (1986). Health and usage monitoring could keep helicopters flying safely. *Engineer*, **263**, 36–37.

Veluswami M. A., Ratnam C. H. *et al.* (1994). An investigation of impact phenomenon in parallel-shaft geared system. In *Proc. COMADEM'94*, Sept.

Wallo M. J. (1984). Microcomputer-based instruments: speed dynamic balancing to minimise stem turbine vibrations. *Mech. Eng.*, **106**, 58–61.

Wallo M. J. and Stot H. R. (1989). Monitoring and diagnostic program for major rotating machinery of an electrical utility. *Mater. Eval.*, **47**, 604–8.

Ward K. A. (1994). Retrofitting old turbomachinery with vibration monitors. *Hydrocarbon Processing*, **73**, 49–51.

Watson W. D. (1988). Vibration analysis key to compressor foundation analysis. *Oil & Gas J.*, **86**, 40.

Watkins K. and Watton J. (1983). Wear monitoring of positive displacement vane pump. In *Proc. 2nd. Int. Conf. on Condition Monitoring*, London, May.

Wilson B. K. (1994). Synchronous averaging analysis of diesel engine turbocharger vibration. *Sound & Vibration*, **28**, 16–21.

Yoshimura H., Ueda A. and Morita M. (1992). Acoustic emission monitoring on a model field winding, etc. *Cryogenics*, **32**(5), 502–7.

Zhuge Q. and Lu Y. (1991). Signature analysis for reciprocating machinery with adaptive signal processing. *Proc. IMechE., Pt. C*, **205**, 305–10.

Zhuge Q. and Yangxiang L. (1991). Vibration source transmission path response analysis and condition monitoring of hydraulic pumps. *J. Fluid Control*, **21**(1), 61–69.

Bearings and gears

Aatola S. and Leskinen R. (1990). Cepstrum analysis predicts gearbox failure. *Noise Control Eng.*, **34**, 53–59.

Akamatsu Y. *et al.* (1992). Influence of surface roughness skewness on rolling contact fatigue life. *Tribology Trans.*, **35**, 745–49.

Amirouche F. M. L. *et al.* (1992). Dynamic analysis of flexible gear-trains/transmissions: an automated approach. *J. Appl. Mech.*, **59**, 976–82.

Antosiewicz M. B. (1992). Predictive maintenance: monitoring the health of gear drives. *Tappi J.*, **75**, 67–68.

Arakere N. K. and Nelson H. D. (1988). Collocation method for finite-length squeeze-film dampers with variable clearance. *J. Tribology*, **110**, 685–92.

Aramaki H. *et al.* (1992). Film thickness, friction and scuffing failure of rib/roller end contacts in cylindrical roller bearings. *J. Tribology*, **114**, 311–16.

Bagnoli S., Capitani R. *et al.* (1988). Accelerometer and acoustic emission signals as diagnostic tool in assessing bearing damage. In *Proc. 2nd. Int. Conf. on Condition Monitoring*, London, May.

Baillie D. C. and Mathew J. (1994). Fault diagnosis of bearings using short data lengths. In *Proc. COMADEM'94*, Sept.

Bandrowski J. C. (1993). Maintaining linear ball-bearings. *Plant Eng.*, **47**, 153–55.

Barry J. E. (1991). How to track rolling element bearing health with vibration signature analysis. *Sound & Vibration*, **25**, 24–35.

Blankenship G. W. and Singh R. (1992). New rating indices for gear noise based upon vibroacoustic measurements. *Noise Control Eng. J.*, **38**, 81–92.

Boyd L. S. and Pike J. (1989). Epicyclic gear dynamics. *AIAA J.*, **27**, 603–9.

Braza J. F. (1992). Rolling contact fatigue and sliding wear performance of ferritic nitrocarburised M50 steel. *Tribology Trans.*, **35**, 89–97.

Campbell C. M. and Tavakoll M. (1993). Comparative signature analysis using needle bearings. In *Proc. 14th. ASME Design Tech. Conf. on Mechanical Vibration and Noise*, Albuquerque NM, Sept.

Carletti E. and Vecchi I. (1990). Acoustical control of external gear pumps by intensity measuring techniques. *Noise Control Eng. J.*, **35**, 53–59.

Chambers W. S. and Mikula A. M. (1988). Operational data for a large vertical thrust bearing in a pumped storage application. *Tribology Trans.*, **31**, 61–65.

Choy F. K., Padovan J. *et al.* (1992). Coupling of rotor gear casing vibrations during extreme operating events. *J. Pressure Vessel Tech.*, **114**, 464–71.

Coe H. H .and Zaretsky E. V. (1987). Effect of interference fits on roller bearing fatigue life. *ASLE Trans.*, **30**, 131–40.

Darling J. and Mu C. (1991). The condition monitoring of rolling element bearings using acoustic emission. In *Proc. 8th World Conf. on the Theory of Machines and Manufacturing, Prague*.

Dimarogonas A. D. (1988). Limit cycles for pad bearing under fluid excitation. *Tribology Trans.*, **31**, 66–70.

El-Saeidy F. M. A. (1991). Effect of tooth backlash and ball-bearing deadband clearance on vibration spectrum in spur gearboxes. *J. Acoust. Soc. Amer.*, **89**, 2766–73.

El Sherif A. H. (1994). Analyse bearing problems by ball path inspection. *Hydrocarbon Processing*, **73**, 120–21.

Georgeson J. D. and Lieu D. K. (1992). Inspection of roller bearing surfaces with laser Doppler vibrometry. *J. Eng. for Industry*, **114**, 123–25.

Gesdorf E. J. (1984). How safe is your lubrication monitoring equipment. *Lubrication Eng.*, **40**, 335–43.

Gupta P. K. and Tallian T. E. (1990). Roller bearing life prediction/correction for materials and operating conditions: implementation in bearing dynamics computer codes. *J. Tribology*, **112**, 23–26.

Hackett W. L. (1990). Optimising the large-scale frequency analysis of paper machine bearings. *Tappi J.*, **73**, 97–104.

Harker R. and Sandy J. L. (1989). Rolling element bearing monitoring and diagnostic techniques. *J. Eng. for Gas Turbines & Power*, **111**, 251–56.

Harker R. G. and Hansen J. S. (1985). Rolling element bearing monitoring using high-gain eddy-current transducers. *J. Eng. for Gas Turbines and Power*, **107**, 160–64.

Harris T. A. *et al.* (1992). The effect of hoop and material residual stresses on the fatigue life of high-speed rolling bearings. *Tribology Trans.*, **35**, 194–98.

Hashimoto H., Wada S. *et al.* (1986). Performance characteristics of worn journal bearings in both laminar and turbulent regimes. *ASLE Trans.*, **29**, 565–77.

Herbage B. S. (1993). Bearing upgrade reduces vibration and improves unit load. *Power Eng.*, **97**, 41–43.

Hills P. W. (1994). Advances in low-frequency data analysis and early bearing/gear wear detection. In *Proc. COMADEM'94*, Sept.

Holzhauer W. (1991). Surface changes around large raceway indentations during run-in of tapered roller bearings. *Tribology Trans.*, **34**, 361–68.

Kishor B. and Gupta S. K. (1989). On dynamic gear tooth loading due to coupled torsional/lateral vibrations in geared rotor hydrodynamic bearing system. *J. Tribology*, **111**, 418–25.

Kleinlein E. (1992). Grease test system for improved life of ball and roller bearings. *Lubrication Eng.*, **48**, 916–22.

Komatsuzaki S. *et al.* (1994). Estimation of service life of grease in large-size roller bearings. *Lubrication Eng.*, **50**, 25–29.

Krishnappa G. (1984). Noise and vibration measurements of 50 kW vertical axis wind turbine gearbox. *Noise Control Eng. J.*, **22**, 18–24.

Li C. and Wu S. M. (1989). Online detection of localised defects in bearings by pattern recognition analysis. *J. Eng. for Industry*, **111**, 331–36.

Lim T. C. and Singh R. (1991). Statistical energy analysis of a gearbox with emphasis on the bearing path. *Noise Control Eng. J.*, **37**, 63–69.

Logan D. B. and Mathew J. (1994). Using the correlation dimension to detect rolling element bearing faults. In *Proc. COMADEM'94*, Sept.

Lorosch H. K. (1985). Research on longer life for rolling element bearings. *Lubrication Eng.*, **41**, 37–43.

Lovell M. R. *et al.* (1993). The response of balls undergoing oscillatory motion: crossing from boundary to mixed lubrication regimes. *J. Tribology*, **115**, 261–66.

Luming H. (1986). Vibration monitoring of paper machine bearings. *Tappi J.*, **69**, 234–36.

Luo M. and Kuhnell B. T. (1991). Forecasting machine condition using grey-system theory. *Cond. Mont. Diagn. Tech.*, **1**(3), 102–5.

McFadden P. D. and Smith J. D. (1984). Acoustic emission transducers for the vibration monitoring of bearings at low speeds. *Proc. IMechE., Pt. C*, **198**(8), 127–30.

McFadden P. D. and Smith J. D. (1985). A signal processing technique for detecting local defects in a gear from the signal average of the vibration. *Proc. IMechE., Pt. C*, **199**(4), 287–92.

McFadden P. D. and Smith J. D. (1985). An explanation for the asymmetry of the modulation sidebands about the tooth meshing frequency in epicyclic gear vibration. *Proc. IMechE., Pt. C*, **199**(1), 65–70.

McMahon S. W. (1991). Condition monitoring of bearings using ESP. *Cond. Mon. Diag. Tech.*, **2**(1), 21–25.

Mark W. D. (1987). The role of the discrete Fourier transform in the contribution to gear transmission error spectra from tooth spacing errors. *Proc. IMechE., Pt. C*, **201**(3), 227–29.

Mark W. D. (1992). Contributions to the vibratory excitation of gear systems from periodic undulations on tooth running surfaces. *J. Acoust. Soc. Amer.*, **91**, 166–86.

Mathew J. and Stecki J. S. (1987). Comparison of vibration and direct reading ferrographic techniques in application to high-speed gears operating under steady and varying load conditions. *Lubrication Eng.*, **43**, 646–53.

Milne R., Aylett J., McMahon S. and Scott T. (1991). Portable bearing diagnostics using enveloping and expert systems. In *Proc. 3rd Int. Cong. on Condition Monitoring and Diagnostic Engineering*.

Nowicki R., Slowinski R. *et al.* (1992). Evaluation of vibroacoustic diagnostic symptoms by means of rough sets theory. *Computers in Industry*, **20**, 141–52.

Pandit S. M. *et al.* (1993). Bearing defect detection using data-dependent systems and wavelet methods. In *Proc. Symp. on Mechatronics, ASME Dyn. Sys. Cont. Div. Publ.*, **50**, 285–93.

Phillips G. J. (1989). Measuring the size of rolling element bearing flaws. *Sound & Vibration*, **23**, 6.

Prashad H. *et al.* (1985). Diagnostic monitoring of rolling element bearings by high-frequency resonance technique. *ASLE Trans.*, **28**, 439–47.

Prashad H. (1987). The effect of cage and roller slip on the measured frequency response of roller element bearings. *ASLE Trans.*, **30**, 360–67.

Prashad H. (1992). Theoretical analysis of capacitive effect of roller bearings on repeated starts and stops of a machine operating under the influence of shaft voltages. *J. Tribology*, **114**, 818–22.

Rahnejat H. and Gohar R. (1985). The vibrations of radial ball-bearings. *Proc. IMechE., Pt. C*, **199**(3), 181–93.

Ratcliffe G. A. (1990). Condition monitoring of rolling element bearings using the envelope technique. *IMechE Publ. on Machine Condition Monitoring*, 55–65.

Rees F. (1987). Monitoring system to predict machinery failure. *Information & Software Tech.*, **29**, 6–7.

Sandy J. (1988). Monitoring and diagnostics for roller element bearings. *Sound & Vibration*, **22**, 16–20.

Savaskan T., Laufer E. E. *et al.* (1984). Wear in a high-speed roller bearing. *Metals Tectn.*, **11**, 530–34.

Sayles R. S. and Ioannides E. (1988). Debris damage in rolling bearings and its effects on fatigue life. *J. Tribology*, **110**, 26–31.

Schlitz R. L. (1990). Forcing frequency identification of rolling element bearings. *Sound & Vibration*, **24**, 16–19.

Schrader S. M. (1992). Performance of a hybrid cylindrical roller bearings. *Lubrication Eng.*, **48**, 665–72.

Smith J. D. (1993). Generation of Smith shocks in gears during oil starvation. *Proc. IMechE., Pt. C*, **207**(4), 279–85.

Smith J. D. (1993). Monitoring the running-in of gears using Smith shocks. *Proc. IMechE., Pt. C*, **207**(5), 315–23.

Stockman S. and Frick N. (1988). Digital data system improves bearing monitoring. *Power Eng.*, **92** 22–23.

Stockman S. and Frick N. (1991). Monitoring system helps hydro plant avoid bearing repairs downtime. *I&CS*, **64**, 104.

Su Y. T., Sheen Y. T. *et al.* (1992). Signature analysis of roller bearing vibrations: lubrication effects. *Proc. IMechE., Pt. C*, **206**, 193–202.

Su Y. T., Sheen T. Y. *et al.* (1993). On the detectability of roller bearing damage by frequency analysis. *Proc. IMechE., Pt. C*, **207**, 23–32.

Sundberg A. (1984). Monitoring bearings with the shock pulse method. *PIMA Mag.*, **66**, 57–59.

Tandon N. and Nakra B. C. (1992). Comparison of vibration and acoustic measurement techniques for the condition monitoring of rolling element bearings. *Tribology Int.*, **25**(3), 205–12.

Timmermann D. N. (1986). Checking the condition of gear drives: diagnosing temperature and noise problems. *Plant Eng.*, **40**, 72–75.

Venalainen E. (1989). Continuous electronic monitoring of oil flows, bearing oil temperatures and the parameters of machinery. *Tappi J.*, **72**, 65–71.

Walford T. L. and Stone B. J. (1983). The sources of damping in rolling element bearings under oscillating conditions. *Proc. IMechE., Pt. C*, **197**, 225–32.

Wardle F. P. (1988). Vibration forces produced by waviness of the rolling surfaces of thrust-loaded ball-bearings. *Proc. IMechE., Pt. C*, **202**(5), 305–12.

White G. (1991). Amplitude modulation: a new tool for predictive maintenance. *Sound & Vibration*, **25**, 14–19.

Worden K., Staszewski W. J. *et al.* (1994). Gear fault detection and severity classification using neural networks. In *Proc. COMADEM'94*, Sept.

Textbooks, standards and company handbooks

Textbooks

Anderson R. T. and Neri L. (1990). *Reliability Centred Maintenance: Management and Engineering Methods*. London: Elsevier.

Braun S. (1986). *Mechanical Signature Analysis: Theory and Applications*. London: Academic Press.

Cempel C. (1991). *Vibroacoustic Condition Monitoring.* Chichester: Ellis Horwood.

Hope, A. D. (1990). *Condition Monitoring Technology: Principles and Practices.* London: Longman.

Hunt T. (1993). *Handbook of Wear Debris Analysis and Particle Detection in Liquids.* London: Elsevier.

Ljung L. (1985). *Theory and Practice of Recursive Identification.* Cambridge MA: MIT Press.

Miller R. K. and McIntire P. (1982). *NDT Handbook, Vol. 5:* New York: *American Society of Metals. Acoustic Emission Testing.*

Pallas-Areny R. and Webster J. G. (1991). *Sensors and Signal Conditioning.* New York: Wiley.

Patton R., Frank P. *et al.* (1989). *Fault Diagnosis in Dynamic Systems: Theory and Application.* Englewood Cliffs NJ: Prentice Hall.

Smith J. D. (1983). *Gears and Their Vibration: A Basic Approach to Understanding Gear Noise.* New York: Marcel Dekker.

Watton J. (1992). *Condition Monitoring and Fault Diagnosis in Fluid Power Systems.* Chichester: Ellis Horwood.

Wowk V. (1991). *Machinery Vibration: Measurement and Analysis,* New York: McGraw-Hill.

Standards

American National Standards Institute (1975). *Balance Quality Requirements of Rigid Rotors (ANSI S2.19-1975).* New York: ANSI.

American National Standards Institute (1984). *Mechanical Vibration of Rotating and Reciprocating Machinery: Requirements for Instruments for Measuring Vibration Severity (ANSI S2.40-1984; revised 1990).* New York: ANSI.

American National Standards Institute (1984). *Criteria for Evaluating Flexible Rotor Balance (ANSI S2.43-1984; revised 1990).* New York: ANSI.

American National Standards Institute (1985). *Mechanical Vibration of Large Rotating Machines with Speed Range from 10 to 200 Rev/Second: Measurement and Vibration Severity In Situ (ANSI S2.41-1985; revised 1990).* New York: ANSI.

British Standards Institution (1987). *Balance Quality Requirements of Rigid Rotors (BS 6861).* London: BSI.

Institute of Electrical and Electronics Engineers (1983). *IEEE Guide to the Collection and Presentation of Electrical, Electronic Sensing Component and Mechanical Equipment Reliability Data for Nuclear Power Generating Stations (STD-500-1984).* New York: IEEE.

International Organisation for Standardisation (1973). *Balance Quality of Rotating Bodies (ISO R 1940-1973)*. Geneva: ISO.

International Organisation for Standardisation (1986). *Balance Quality Requirements of Rigid Rotors (ISO 1940/1)*. Geneva: ISO.

Verein Deutsche Ingenieure (1979). *Balance Quality Requirements of Rigid Rotors (VDI 2060)*. Berlin: VDI.

Company Handbooks

Angelo M. (1987). Vibration monitoring of machines. *Bruel & Kjaer Tech. Rev.*, No. 1.

Bernhard D. L. (1993). The Practical Application of ISO 1940/1: Balance Quality Requirements of Rigid Rotors. *Technical Paper 14*, IRD Mechanalysis Inc.

Bradley D. (1993). *Windows to Quality Maintenance Reporting*. IRD Mechanalysis Inc.

Broch J. T. (1980). *Mechanical Vibration and Shock Measurements*. Bruel & Kjaer.

Courtney S. (1993). *Vibration Based Predictive Maintenance: The Training Requirement*. IRD Mechanalysis Inc.

Courtney S. (1992). *IRD Fast Track Spike Energy Spectrum*. IRD Mechanalysis Inc.

Diagnostic Instruments (1994). *Machinery Condition Monitoring and Vibration Analysis*. Diagnostic Instruments Ltd.

Endevco (1994). *Portable Condition Monitoring Systems*. Endevco UK Ltd.

Findley F. A. (1993). *A Systemic Approach to Solving Vibration Problems*. IRD Mechanalysis Inc.

Holroyd T. J. and Randall N. (1994). *Field Application of Acoustic Emission to Machinery Condition Monitoring*. Holroyd Instruments.

Holroyd T. J. and Randall N. (1994). *Applying Acoustic Emission to Industrial Machinery Monitoring*. Holroyd Instruments.

Holroyd T. J. and Randall N. (1994). *Cost-effective Condition Monitoring Based on Acoustic Emission*. Holroyd Instruments.

Howieson D. D. (1991). *A Practical Introduction to Condition Monitoring of Rolling Element Bearings Using Envelope Signal Processing*. Diagnostic Instruments.

IOtech (1991). *Instrument Communcation Handbook: Interfacing Standards*. IOtech Inc.

IRD Mechanalysis (1988). *Vibration Technology 1: Student Handbook*. IRD Mechanalysis Inc.

IRD Mechanalysis (1990). *Dynamic Balancing Handbook*. IRD Mechanalysis Inc.

March R. (1994). *Cost-Benefit Analysis Methods for Condition Monitoring*. Solartron Instruments.

Mellor D. J. (1994). *Rolling Element Bearing Damage Detection Using the Kurtosis Technique*. Condition Monitoring Ltd.

Mellor D. J. (1994). *Spectrum Monitoring: An On-Line Approach*, Condition Monitoring Ltd.

Randall R. B. (1981). Cepstrum analysis. *Bruel & Kjaer Tech. Rev.*, No. 3.

Setford G. A. W. (1992). *Bearings, Condition Monitoring, Condition Measurement and Condition Control*, SPM Instruments.

Shreve D. H. (1993). *Helping the Company Bottomline: Predictive Maintenance Programs are Key.* IRD Mechanalysis Inc.

Sohoel E. O. (1984). *Shock Pulses as a Measure of the Lubricant Film Thickness in Rolling Element Bearings*. SPM Instruments.

Index